Synthese Library

Studies in Epistemology, Logic, Methodology, and Philosophy of Science

Volume 380

More information about this series at http://www.springer.com/series/6607

Acknowledgments

I've been working on this project for a long time. Many people have helped along the way. Among them: Jamin Asay, Alejandro Bárcenas, Karen Bennett, Stephen Biggs, Valerie Brown, Otávio Bueno, Brian Casas, Walter Edelberg, Bucky Farley, Sam Fleischacker, Eric Gilbertson, Craig Hanks, Rebecca Hanrahan, Dave Hilbert, Bill Hart, Peter Hylton, Charlotte Jackson, Colin Klein, Felipe Leon, Vince Luizzi, Chris McNiven, Josh Norton, Burkay Ozturk, Bev Pairett, Sònia Roca-Royes, Marya Schechtman, Scott Shalkowski, Todd Stewart, Daniel Sutherland, John van Dyke, Anand Vaidya, Peter van Inwagen, Aleks Zarnitsyn, and audiences at far too many conferences. I'm grateful to you all.

Parts of this book are based on parts of the following essays: "Amie Thomasson's *Ordinary Objects*," *Metaphilosophy* (2009); "Modal Knowledge, in Theory," *Southwest Philosophy Review* (2012); "Theory Selection in Modal Epistemology," *American Philosophical Quarterly* (2015); "The Modal-Knowno Problem," *Southwest Philosophy Review* (2016) (coauthored with Felipe Leon, though I only used the portions I wrote); "Modal Empiricism: Objection, Reply, Proposal," in *Modal Epistemology After Rationalism* (Synthese Library), edited by Bob Fischer and Felipe Leon (Dordrecht: Springer, 2016); and "A Theory-Based Epistemology of Modality," *Canadian Journal of Philosophy* (2016). Thanks to the publishers for allowing me to rework and integrate all this material.

Finally, a note to my children. Despite having labored over this book for more than 10 years, I'm not satisfied with what I've done. In fact, I can only see the flaws and limitations of my work. In this respect—as in many others—don't imitate me. Instead, learn to see the good in what you do. I promise you: it's there.

For Anand and Felipe, who encouraged me to keep going.

Bob Fischer

Modal Justification via Theories

Springer

Bob Fischer
Department of Philosophy
Texas State University
San Marcos, TX, USA

Synthese Library
ISBN 978-3-319-49126-4 ISBN 978-3-319-49127-1 (eBook)
DOI 10.1007/978-3-319-49127-1

Library of Congress Control Number: 2016961292

Printed on acid-free paper

This Springer imprint is published by Springer Nature
The registered company is Springer International Publishing AG
The registered company address is: Gewerbestrasse 11, 6330 Cham, Switzerland

Contents

Part II

Part I

Chapter 1
A Theory-Based Epistemology of Modality

Abstract This book articulates and defends a *theory-based epistemology of modality* (TEM). At TEM's core is the idea that if we're justified in believing the extraordinary modal claims to which philosophers often appeal—such as the claim that I could be disembodied, or that there could be a maximally perfect being—it's because we're justified in believing theories according to which those claims are true. This chapter articulates some of the assumptions that shape TEM, sketches the view, and then develops the basic argument for it, which is based on an analogy between a plausible modal epistemology for games and modal epistemology generally.

This book articulates and defends a *theory-based epistemology of modality*—henceforth, TEM. At TEM's core is the idea that if we're justified in believing the extraordinary modal claims to which philosophers often appeal—such as the claim that I could be disembodied, or that there could be a maximally perfect being—it's because we're justified in believing theories according to which those claims are true. That is, TEM says that theories come first, epistemically-speaking. If we're justified in believing an extraordinary modal claim, the ultimate explanation of that fact shouldn't be in terms of, say, that claim's conceivability. Rather, it should be in terms of the argument we offer for a theory—in particular, one according to which that modal claim is true.

This book has two parts. In the first, I provide the basic argument for TEM, which is based on an analogy with games. Then, I provide the details of the view and defend it against a number of objections. In the second part, I consider how TEM fares relative to some general desiderata for choosing among modal epistemologies. I there concede that TEM is a *cautious* modal epistemology—i.e., it makes it harder to maintain that we're justified in believing the extraordinary modal claims to which philosophers like to appeal. This is *not* because TEM involves a thesis to the effect that philosophers aren't (or anyone else isn't) justified in believing particular theories. Rather, it's because philosophers tend to trust certain mental operations—such as conceiving, imagining, and intuiting—even when those operations aren't informed by underlying, justified theory. According to TEM, such trust is misplaced. With this in mind, we might wonder about the extent of TEM's caution. So, I spend

B. Fischer, *Modal Justification via Theories*, Synthese Library 380,
DOI 10.1007/978-3-319-49127-1_1

a chapter clarifying TEM's relationship to van Inwagen's (1998) modal skepticism, arguing that while it offers a better basis for that position than the one van Inwagen provides, it's possible to generate a non-skeptical version of TEM. The bar, however, is high. This invites the objection that TEM is worse than its competitors in terms of its ability to capture our pre-theoretic judgments about those beliefs that are and aren't justified. Against this objection, I argue that some prominent modal epistemologies aren't any less cautious, whatever their proponents might claim for them. If I'm right, being cautious doesn't count too heavily against TEM, if it counts against it at all.

TEM is a new account of how we can be justified in believing certain modal claims. However, there's a sense in which it needn't compete with every other modal epistemology. Consider, for example, Williamson's (2007) counterfactual approach, which explains our modal knowledge by appealing to our ability to evaluate these conditionals while holding fixed certain constitutive principles. As Roca-Royes (2011) argues, Williamson has little to say about how we come to know those constitutive principles. If he were so inclined, he could adopt TEM as his story about how we come to justifiably believe them, and then take the process of evaluating counterfactuals to be his story about the mechanism by which we transfer justification from our theories to specific modal claims. Likewise, think about the views of Lowe (2012) and Hale (2003, 2013), who base our modal knowledge on knowledge of essence. Both Lowe and Hale are quite clear on how we derive the former from the latter, but not nearly as clear about how we get knowledge of essence in the first place. They too could address this issue with TEM, where knowledge of essence emerges from our theorizing about the entities in question. Related points apply to Peacocke's (1999) "principles of possibility" account, Hart (1988) and Kung's (2010) imagination-based views, and the argument-based approach that I sketched in my (2015). TEM is an account of the source of our justification, and it's compatible with different stories both about which mental operation is best suited to producing modal justification and about how modal information should be organized (i.e., in terms of essences or not). In many cases, TEM doesn't have to be a rival. Instead, it can be a supplement.

But even if it doesn't need to be a rival, TEM is one at present. Each of these theories—and all the others of which I'm aware—disagree with TEM about the source of our modal justification. Williamson locates in our ability to assess counterfactuals, our knowledge of constitutive principles, and certain logical equivalencies; Lowe and Hale place it in our ability to divine the essences of things; Peacocke sees it in our coming to know certain general propositions; and so on. By contrast, TEM builds our modal justification on the back of our best theories and the arguments we offer for them. This book explains how and why.

The rest of this chapter provides the groundwork for Part I. I begin by articulating some of the assumptions that shape TEM, sketch the view, and then offer the basic argument for it.

1.1 Some Assumptions

We all take some things for granted when we pursue our philosophical projects. However, we aren't always open about what they are. This brief section is an attempt at transparency: it should help to clarify why I develop TEM as I do. To that end, I'll identify four assumptions.

First, I assume scientific realism—the reason for this will be plain in the next chapter. Scientific realism has a number of facets, but the one relevant here is that we can be justified in believing that our best theories are approximately true. I have nothing novel to say about arguments for scientific realism, the nature of theory confirmation, or how to understand approximate truth. Hence, I won't be discussing these issues. I mention them for two reasons: first, to acknowledge their complexity; second, to set them aside.

Second, consider the claim that an object could travel four billion meters per second. I assume this claim is true or false *mind-independently*. That is, I assume that it's mind-independently true or false in whatever sense it's mind-independently true or false that I had tea before breakfast on May 4, 2016, and that hydrogen has an atomic weight of 1.008, and that $2 + 2 = 4$. I simply ignore all forms of anti-realism about modality. They don't get a hearing in this book.

These first two assumptions are relatively straightforward. The third and fourth require a bit more discussion.

The third assumption is that although the various non-realist alternatives are not in view (conventionalism, fictionalism, expressivism, etc.), I don't assume any particular form of realism about modality (always to be distinguished from Lewis's modal realism, which is only one type of realism about modality). The goal here is to articulate and defend a modal epistemology that's available to all realists, regardless of their metaphysical commitments.

I should be clearer about the significance of this third assumption. It's long been thought that the epistemology and metaphysics of modality ought to go hand in hand. In a sense, that thought goes back to Benacerraf's Dilemma—a familiar problem in the philosophy of mathematics. Benacerraf (1973) argues that the semantics and the epistemology of mathematics pull in two different directions: the platonist offers a semantics for mathematics that's in line with the one that we prefer elsewhere, but also posits entities with which we can't interact; the logicist has a straightforward epistemology, but gets it by way of an inadequate account of mathematical truth. There is an analogous dilemma for modality. Realists about modality offer the most plausible semantics for modal discourse (or so I'll suppose), but they invoke entities with which we can have no causal contact, whether abstracta or causally-isolated concreta. Antirealists can offer straightforward epistemologies, but they manage this only by analyzing modal claims in ways that undermine much of their interest. In mathematics and modality, the demands of truth don't seem to match the demands of knowledge.

Given my second assumption—namely, that modal claims are true or false in a mind-independent way—let's focus on the realist horn of the dilemma. Benacerraf generates the problem by appealing to a causal theory of knowledge, which is no longer in vogue. But we can generate the issue without it. In short, causal contact helps *explain* epistemic success when it occurs, and absent such an explanation, we begin to doubt whether our epistemic successes deserve the name. Without the ability to appeal to causation, we're forced to look for an alternative theory of our accuracy regarding modal matters. Since such theories are few and far between, many philosophers have opted to tinker with their account of modality, somehow bringing us into contact—directly or indirectly—with the truthmakers for the relevant claims. Armstrong (1989), Koons (2000), Rea (2002), Molnar and Mumford (2003), and O'Connor (2008) all are, in one way or another, examples of this reaction.[1] These philosophers are determined to answer "the Integration Challenge," which is Peacocke's (1999) term for the task of making your metaphysics and epistemology fit together.

In the last 40 years, though, the vast majority of those who have written about modal epistemology haven't worried about the Integration Challenge. Taking some form of realism about modality for granted, they've simply proceeded with their epistemological projects. They seem to think that we can do a lot of theorizing about the sources and scope of our modal knowledge without being able to explain our epistemic successes. I assume that this approach is legitimate. So, when I say that my goal is to develop a modal epistemology that's available to all realists, I'm really assuming that we can be confident enough in some epistemic claims to theorize about modal knowledge without devoting much thought to the metaphysics of modality, or how exactly our epistemology fits with it. There is indeed an Integration Challenge, but I devote little attention to it here.[2]

Fourth and finally, I assume that when it comes to the core questions in the epistemology of modality, little turns on the difference between justification and knowledge. I think there are historical reasons for this assumption. Perhaps there are some philosophers who are currently working on the epistemology of modality who didn't first care about the metaphysics of mind, or personal identity, or the philosophy or religion, or what have you. However, the field was born out of debates about particular modal claims—most famously, Descartes's suggestion that you could exist without your body (later refined into the claim that there could be pain without c-fibers firing—see Kripke (1980)). This shaped the conversation around two questions. First, what's the *scope* of our modal knowledge? That is, how *much* do we know about modal matters? Second, what's the *source* of our modal knowledge? *How*, exactly, do we know what we know? In the context of debates about dualism, conceding that your interlocutor was justified was tantamount to conceding the argument entirely. Hence, few (if any) worried whether modal beliefs meet the conditions for knowledge *given* that they're true and justified. I'm working

[1]Of course, others have become non-realists on epistemic grounds—e.g., Sidelle (1989).

[2]I say a bit more elsewhere: see my (2016) and (Forthcoming).

in this tradition, which means that I too am not particularly concerned about the difference between justification and knowledge. Strictly speaking, then, although TEM is a hypothesis about when an agent is *justified* in believing a certain sort of modal claim, you shouldn't make too much of this fact. I'm open to the stronger version of the view—i.e., you *know* a modal claim only if you know a theory according to which it's true—and I'll sometimes slide between talk of knowledge and justification. Unless the context makes it clear that the difference matters, it's safe to assume that it doesn't (or, at least, that it seems to me that it doesn't).

1.2 TEM: A Sketch

With these preliminaries behind us, we can now formulate TEM a bit more clearly. To this end, it will help to get a handle on van Inwagen's (1998) distinction between ordinary and extraordinary modal claims. He suggests a few ways of drawing the distinction, which I'll discuss in Chap. 5.[3] For now, though, we can get by with some of his examples:

Ordinary

- The table that was in a certain position at noon could have been two feet to the left of where it in fact was.
- The legs and top of this table might never have been joined to one another.
- John F. Kennedy could have died of natural causes.
- There could be a full-scale papier-mâché mock-up of a barn that looked like a real barn from a distance.

Extraordinary

- There could be a pure, phenomenal color in addition to red, yellow, green, and blue.
- There could be a three-inch-thick sheet of solid iron that's transparent to visible light.
- There could be a naturally purple cow.
- It's possible for there to be a perfect being
- It's possible that I exist and nothing material exist.
- It's possible that there exist vast amounts of suffering for which there is no explanation.

Van Inwagen's *ordinary* modal claims concern entirely familiar things, such as the locations and features of material objects, and the trajectories of human lives, as well as some unfamiliar things things—i.e., a full-scale papier-mâché mock-up of

---

[3]More precisely, he distinguishes between ordinary claims that those that are "remote from the concerns of everyday life" (1998, 76), but it's become customary to refer to the latter as extraordinary.

a barn. However, it's worth noting that you probably have the skills required to *build* the unfamiliar thing, and you can imagine doing so considerable detail. By contrast, *extraordinary* modal claims concern *entirely* unfamiliar things, and you almost certainly lack the skills to build three-inch-thick sheet of transparent iron. Although it's difficult to give necessary and sufficient conditions for being on either list, I suspect that any reader of this book could extend both of them in ways that van Inwagen would find perfectly acceptable.

However, there are sure to be some gray cases. Where should we place *technological* modal claims—e.g., "It's possible to make a phone that's only 2 mm thick"? On the ordinary / extraordinary spectrum, this is bound to be closer to the ordinary end. However, I certainly couldn't make a phone like that (which isn't saying much, as I couldn't build any such device), and I don't really understand the physical factors that limit the thinness of phones. Likewise, where should we place *scientific* modal claims—e.g., "It's possible for a liquid to have the properties of a gas"? As it happens, I know that this is possible because I've read about such liquids. If you increase the temperature enough, you can convert a liquid into a gas. But if you increase the temperature and pressure enough simultaneously, you can create a *supercritical* liquid, which can do all sorts of interesting things, like pass through a solid as a gas would. However, *prior* to reading about such liquids, I wouldn't have had any idea whether they were possible. For that reason, claims like this seem to be much closer to the extraordinary end of the spectrum.

We aren't going to eliminate the vagueness here, but we can do better than the ordinary / extraordinary distinction. Let's say that modal claims are "interesting" if they aren't completely mundane—i.e., they are concern what's technologically possible, or what's possible according to our best scientific understanding of the world, or what's metaphysically possible. Their interest lies in two facts: (a) that they sometimes matter for philosophical purposes and (b) that we sometimes disagree about them, whereas we tend not to disagree about whether I could have worn different shoes today. In any case, van Inwagen would classify some of the interesting claims as extraordinary, but probably not all, and we won't worry about which are which. When I talk about "interesting" modal claims, I'll be talking about this wider range; when I talk about "extraordinary" modal claims, I'll be talking about modal claims like those that van Inwagen gives as examples, postponing questions about what they might (or might not) have in common.

In any case, with the notion of interesting modal claims in hand, we can formulate TEM. TEM isn't an account of what it takes to be justified in believing *every* modal claim: instead, it's a theory about what it takes to be justified in believing *interesting* modal claims. TEM says that we are justified in believing any interesting modal claim *p* if and only if (a) we justifiably believe a theory according to which *p* is true, (b) we believe *p* on the basis of that theory, and (c) we have no defeaters for the belief that *p*.

I'll say more about this restriction in the next two sections. For now, let's just note that TEM is an account of the *source* of our justification for believing interesting modal claims. There are two implications of this. First, TEM doesn't try to be an account of every instance in which an agent is justified in believing an interesting modal claim. Instead, it's an account of *non-derivative* justification—i.e., the basic source of justification, rather than derivative sources like memory or testimony, which preserve or transmit justification that was earned in other ways. So, for example, as long as I stand in the right epistemic relationship to someone who satisfies TEM's conditions, it's perfectly compatible with TEM that I justifiably believe an interesting claim *p* even though I don't justifiably believe any theory according to which *p* is true. But derivative justification isn't my focus here, and I won't say anything more about it.

Second, although I do think that we're justified in believing all sorts of *interesting* modal claims, I don't assume that anyone is justified in believing any *extraordinary* modal claims. Perhaps no one enjoys this status—perhaps we don't or can't satisfy TEM's conditions. In that case, we should be "modal skeptics," which is van Inwagen's term for the view that "we often do know [ordinary, technological, and scientific] modal propositions, ones that are of use to us in everyday life and in science and even in philosophy, but do not and cannot know (at least by the exercise of our own unaided powers) [extraordinary] modal propositions" (1998, 69).[4] Of course, *if* someone is justified in believing a theory according to which it's possible that she exist and nothing material exists, believes as much on the basis of that theory, and has no defeaters for her belief, then she's entitled to the believe that she could exist and nothing material exist. Frankly, I have no idea whether anyone justifiably believes such a theory. But if someone does, then so much for modal skepticism. My goal is to defend a view about the right epistemic standard for interesting modal claims generally, and I leave it open whether anyone meets it when it comes to extraordinary cases.

Vaidya (2015) distinguishes between uniform and non-uniform modal epistemologies. Uniform modal epistemologies posit a single source of our justification concerning modal matters. As you'd guess, non-uniform accounts posit multiple sources. TEM is a non-uniform account: I don't claim that it can explain all modal justification.[5] I allow that some modal claims are Moorean; I grant that we justifiably believe others by making analogical arguments and judgments of relevant similarity; I'll concede—at least in logic and mathematics—that some of our justification is based on conceptual analysis and other a priori methods. I do, however, deny that any interesting modal claim is Moorean; I also deny that we can defend such a claim based on analogical arguments, judgments of relevant similarity, or conceptual

[4]Van Inwagen acknowledges that the name is a poor one, since it suggests that the view is more radical than it is. As a result, some have taken to calling his view "moderate modal skepticism," distinguishing it from the view that we have no modal justification whatever. That's certainly more accurate, but for ease of exposition, I'll use the shorter name.

[5]I do, however, explore how much TEM could explain—see below.

analysis.[6] I'm fine with other routes to uninteresting modal justification; I'm not fine with other routes to interesting modal justification.[7]

Why would someone be inclined toward a view like this? In the rest of this chapter, I try to answer that question.

1.3 A Modal Epistemology for Games

We have some justified beliefs about modal matters. For example: I could have had tea rather than coffee with breakfast this morning; the coin could have come up heads (though, in fact, it came up tails); that support can't hold more than five hundred pounds. A modal epistemology should state the conditions under which these beliefs are justified. Given that we're realists about modality, what sort of explanation should we expect? That is, if we reject *complete* modal skepticism, and further suppose that the modal facts are mind-independent, what form should we expect our modal epistemology to take? I suggest an answer to this question based on an analogy with games—an answer that should shed some light on TEM's appeal.

We begin with *Clue*—that old murder mystery board game. The object of *Clue* is to be the first one to determine who killed Mr. Boddy. The game has six characters, all of whom are suspects. There are six weapons with which Mr. Boddy might have been killed, and nine rooms in which the murder might have occurred. There are no restrictions on how suspects, weapons, and rooms can be combined. So if we're asked at the outset of the game whether Colonel Mustard could have killed Mr. Boddy, we'll have no trouble answering the question: he certainly could have. Moreover, we know how the murder could have occurred: Mustard could have ended Boddy in the Ballroom with the lead pipe, or in the Library with the noose, or whatever. Alternately, suppose we're asked whether Mustard could have killed Boddy in one of the hallways that connect the various rooms. The answer is equally clear: certainly not. The murder can only occur in a space on the board having a corresponding card in the deck, and while there is a card for each room, there isn't a card for any of the hallways. Where *Clue* is concerned, we have modal knowledge in spades.

But consider this: if you didn't know the rules of *Clue*—and if, for some sad reason, no one would share them with you—how could you determine whether Mustard could have killed Boddy? I see only a few options. First, and most

[6]Of course, if we start with an extraordinary modal claim as a premise, then we may be able to infer other extraordinary modal claims from it. (I call this "conditional" modal justification, and discuss it in Sect. 3.3.) But cases like this aren't the interesting ones. What matters is how we get extraordinary modal justification in the first place, not what we can do with it once we've got it.

[7]Expressions like "ordinary modal justification" are shorthand for the less-wieldy "justification for believing ordinary modal claims."

obviously, you could watch game after game until you happen to see one in which Mustard is the guilty party. Second, you could reason analogically or inductively: "I've seen a game in which Professor Plum kills Boddy; Plum seems similar to Mustard; so, Mustard could have killed Boddy too." Finally, you could try to figure out the rules. Abstractly, this would involve postulating different kinds of objects (the board; the characters; the cards; etc.), attributing different properties to the instances of those kinds, and proposing general rules that govern their interactions. Put differently, you could (a) make an inference from actual gameplay to possible gameplay, (b) reason analogically or inductively from cases, or (c) develop a working theory of *Clue*'s gameplay.

Plainly, these three ways of answering your question are related. An inductive argument will presuppose some categorization scheme, as will the observation *that* thus and so, and theory development depends on observation and a range of ampliative inference types. The reason to draw these distinctions isn't to suggest that there are wholly independent sources of evidence. Rather, I want to make two points.

First, while the relationship between these question-answering strategies might be controversial, I hope it's obvious that *they are indeed the strategies*. The reason for this is simple. The modal facts of *Clue* aren't yours to stipulate, and guesses will be just that. So, you have to try to extrapolate the facts in question from the only data available—namely, your observations of gameplay. Second, I hope it's equally clear that, as you consider increasingly odd cases, you'll need to rely more on your theory of *Clue*'s gameplay and less on the other sources of evidence. If you're considering whether Plum can land on a particular space, observation will probably settle the case; if it doesn't, an analogical argument seems fine ("I've seen him land on similar spaces, and there's no obvious reason why this one would be off limits"). But if you're considering whether Plum can enter and exit a room in a single turn, or step off one end of the board to arrive at another, then mere observation won't help. Likewise, there may not be any case that seems to be relevantly similar to the one now in question. Or there might be at least two cases that are similar in important respects, but the analogies point in different directions. Hence, a theory of the rules in your best shot.

In ordinary cases, after all, you can get pretty good evidence that you haven't overlooked a rule forbidding precisely what looks to be allowed. You've seen many character / space combinations already, and the only differences between them and the present character / space combination are the identities of the particular character and particular space. So, after watching the characters move through the hallways for a while, you'd demand a reason to think that Plum *couldn't* land on a space that you've not seen him land on before. Here, you've got no evidence for a rule that precludes the combination, which looks like fallible evidence for saying that there is no such rule. In weirder cases, however—such as entering and exiting a room in a single turn—you can't draw on your experience of similar events. So, the absence of evidence isn't evidence of anything. Your best bet is to see whether the most comprehensive story you can tell about the rules delivers a verdict about the matter at hand. If it does, then your evidence for that story is some evidence for that verdict. If it doesn't, then you've got nothing to go on, and you should suspend belief.

Of course, you've seen all sorts of moves in the past, and you might think that at least some of them are relevant to the in-and-out-in-a-single-turn question. I think so—though not via an analogical argument. Instead, as you build up a theory of the game, you distinguish various kinds of objects—the board as whole, with hallways and room within it; the characters; the weapons; the cards; etc.—and various kinds of moves—walking down hallways, entering rooms, making an accusation, etc. Among other things, these categories help you assess what is and isn't similar for present purposes. Here, relevant cases are those involving (a) entering rooms, (b) exiting rooms, and (c) reversing direction mid-turn. Let's suppose you've seen cases of the first two types (albeit not combined), but never the third. Perhaps the best explanation of *not* seeing direction reversals is that they're forbidden; then, you might infer that entering and exiting in a single turn would be like a direction-reversal, and hence isn't allowed either. But you also have reason to think that room entrances and exits are special cases, since there are other events that occur only immediately following or preceding them, such as offering a theory of the crime. This provides a reason to worry about whether the analogy between entering / exiting and direction-reversals is well-founded. To address this, you might need to draw on a larger story about the point of entering and exiting rooms. The short of it is that your experiences are indeed relevant, but this is largely because they're the basis of a more systematic account of how the game works. So, the better your grip on the rules, the better your odds of assessing what can and can't be done. This is a fallible process; any step in this reasoning might be mistaken. But since there's no alternative, we just have to make the best of it.

1.4 The Game Analogy

We've just outlined a modal epistemology for *Clue*, and I submit that it's a good model for all modal knowledge.[8] Of course, when we compare ordinary modal inquiry to modal inquiry in *Clue*, we can't watch the world unfold all over again. This isn't a problem for the analogy, but a point in favor of epistemic humility. In ordinary modal inquiry, it's harder to tell whether we are being deceived by coincidences. But epistemic humility is fitting. We don't make up the modal facts, nor do we have any faculty of modal intuition that connects us with them. So our best shot is to make inferences from actuality to possibility, to offer analogical and inductive arguments, and to do our best to figure out the world's rules.

After all, the world has rules too: natural, metaphysical, and logical, among others. And like the rules of *Clue*, these rules define possibility spaces. I'm not committing myself to any particular account of these rules. For all I've said here, talk about the world's rules could be a way of referring to the various laws that govern what's possible—e.g., fundamental physical laws for nomic possibility and,

[8]People can cheat in games. But let's assume, just for the sake of simplicity, that we were watching a bunch of rule-followers play *Clue*.

say, the principle of recombination for metaphysical possibility. Or perhaps talk of rules is shorthand for the true generalizations about the countless modal truths that are grounded in the real essences of objects, or shorthand for the true generalizations about primitive modal facts. The details aren't especially important here. All that matters is that there be some mind-independent facts that determine the space of possibilities.

How do we *know* that there are factors that determine the space of possibilities? Frankly, I'm not sure we do. However, if we're realists about modality and there aren't any such factors, then we should despair of having any modal knowledge whatever. If modal reality is not tolerably well-*organized*—if we can't extrapolate some facts about the space of possibilities using principles that hold in the actual world—then we should simply stop talking about the ways that things could and couldn't be. By comparison, imagine if we weren't willing to assume that certain general truths that hold within our lightcone also hold outside it.[9] Given as much, on what basis would we say anything at all about the distant parts of the universe? Surely none. And the same point applies, *mutatis mutandis*, to the idea that we might have some basis for venturing an opinion about a chaotic space of worlds.[10]

In my view, then, we have to choose between (a) making the assumption that there are detectable factors that jointly determine the space of absolute possibilities and (b) denying that we have any modal justification whatever. Whether we're willing to make the assumption will probably come down to our threshold for epistemic risk. But for those of us who can't see how to get by without modality, it will be obvious what to say.

[9]On the assumption that no cause can travel faster than the speed of light, we can say, very roughly, that our light cone is the region of spacetime containing both (a) the events that could causally affect us and (b) the events on which we might have some causal impact. See Huggett (2010) for a philosophically-astute introduction.

[10]Another analogy. Imagine that you're a spy who's captured in the middle of a daring mission. Your captors drug you and take you to an undisclosed location. You wake up slumped in the corner of windowless room; two of your captors are standing in the opposing corner, guns drawn. They say:

> You are in a building. This may or may not be the building's only room. Give us your estimate of the contents of this building, being careful to indicate, for any two objects, whether or not you judge them to be in the same room. And by the way: you may not leave this room, we will not answer any questions about the building or its contents, and you should not assume that the other rooms—if others there are—bear any resemblance to this room.

And of course, because this is a spy story, they go on to say: "If your estimation is largely correct, then we will release you immediately; if it is not, then we will torture you until you give up national security secrets." There should be little doubt that you are about to be tortured. There is virtually *no way* that your estimation will be correct; you have no information from which you can plausibly extrapolate the facts about the building. But now let's suppose that realism about modality is true, and that we are causally isolated from the truthmakers for modal claims, and that the worlds are not organized in any discernable way (i.e., if they have no invariant features that we can discover by theorizing about actuality). If all this is true about modal reality, then what's the difference between your epistemic situation in our spy story and your epistemic situation vis-à-vis modal reality? To my mind, none whatever.

In any case, as with *Clue*, I think we should expect to rely on observation, analogy, and induction when making uninteresting modal judgments. But as we consider interesting modal claims—such as how much weight a bridge could hold, or whether we could resurrect an extinct species via genetic engineering, or whether entangled particles can be in a particular state—we should start looking to our best theories for guidance.

The reason for this is simple. The basic worry when assessing one modal claim is that we're overlooking another with which it's incompatible. When considering uninteresting claims—the furniture could have been arranged otherwise, I could have taken the back way to work, etc.—we have a large reference class of similar cases on which to draw. These cases constrain the sort of modal claim that might threaten the one in question. And the larger the reference class, the less plausible it is that those specific modal claims are true while all the others of the same kind are false. Here, the absence of evidence for the conflicting modal claims are evidence of their absence.

However, we don't have relevant cases on which to draw when we consider many technological and scientific modal claims—that's often why those claims are interesting. So instead, our evidence for our theories is evidence for the absence of conflicting modal claims. If our best biology implies that you can make a dinosaur by inserting certain genes into a chicken egg, then the evidence for that biology is evidence that nothing precludes making such a creature. But if we have no theory that that speaks to the claim in question, and it's a claim that's well beyond our experience, we have little reason to doubt that there's a conflicting modal claim. Here, the absence of evidence for any conflicting modal claims isn't evidence for anything, and we should just admit our ignorance.

The same point applies to philosophically-significant cases, such as disembodied existence, utility monsters, and swamp-persons. Let's focus on the last example. Given what we know about swamps and persons and lightning, it isn't at all clear how lightning could produce a person from a swamp, nor whether that state of affairs is possible. So absent a theory, we don't have any reason to suppose that there isn't a necessity that precludes the existence of swamp-persons, since everyday experience provides no analogous cases from which we can make an inference. So absent a theory, we should be agnostic about whether swamp-persons are possible.

To be clear: I am not arguing for skepticism about extraordinary modal claims, much less skepticism about technological or scientific or metaphysical claims. In principle, we might justifiably believe theories—physical or metaphysical—that speak to all the claims we care about. Suppose, for example, that we justifiably believe physicalism to be true. Then, we can infer that it isn't possible to have mental states without physical states. This is because the evidence for physicalism is evidence for the absence of a conflicting modal claim (e.g., claims that there can be disembodied minds).[11] But if the analogy with *Clue* holds, then absent evidence

[11]Granted, this runs counter to the methodology that metaphysicians often employ. Hart (1988) and Chalmers (1996), for example, offer arguments for dualism that stand or fall on contentious

for such a theory, I don't see why we'd trust any inclination to say that there could, or couldn't, be disembodied minds.

Still, some might be inclined to say that, absent defeaters, we can trust our modal judgments independently of whether we justifiably believe relevant theories. This might be plausible if we had some story about why we're generally reliable about modal matters, so that a presumption of reliability could trump our absence of evidence. Then, if it seemed to us that *p* is possible, we could be justified in believing as much in the absence of defeaters. O'Connor (2008) has a view like this: he develops a kind of theistic reliablism on which God's power grounds the modal facts, and God has constituted us so that (when functioning properly, and we're in the right environment, etc.) we form true beliefs about modal matters. I raise no objection to this story here, but let's note the obvious. First, the account isn't available to most modal epistemologists. Second, God isn't easily replaceable in it; I'm not aware of any impersonal, natural mechanism to which non-theists can appeal to ensure reliability.

At this juncture, someone might want to appeal to natural selection. Granted, it might make us reliable in some cases, though it's certainly a cruder mechanism that divine fine-tuning would be, and you'd think that greater epistemic caution would be appropriate. But more importantly, we should recall that evolutionary forces (a) are constrained by time and the available raw materials and (b) are not concerned, first and foremost, with producing true beliefs (even if true beliefs they produce). These points lead to three serious problems. First, given that evolutionary forces will be happy to trade truth for another benefit if the opportunity presents itself, we need a reason to think that they've operated long enough in conditions favorable to selecting reliable mental mechanisms. (Granted, we have reason to suppose that our mental mechanisms are superior to those had by our competitors, but for all we know, that's a backhanded compliment.) What argument can we offer for this view? Second, given that evolutionary forces are not directed toward truth, it would appear to be an *accident* that our faculties are reliable (if reliable they are). But if it is an accident that our faculties are reliable, we should ask why *this* accident is an epistemic improvement over the seemingly-accidental correlation with which we began—namely, the correlation between modal matters and our beliefs about modal matters. What answer can we offer? Third and finally, even if evolutionary forces have been operating in the right conditions for a sufficient amount of time, and even if our reliability is not an accident (or at least not an epistemically-

---

modal premises that are not supposed to presuppose any particular theory of mind. Indeed, many of us were drawn to the epistemology of modality *precisely because* metaphysicians argued for theories based on extraordinary modal claims, and we wanted to know whether those claims could be justified. That, at least, is my own story. However, the analogy with *Clue* suggests that it's a mistake to argue for theories based on extraordinary modal claims. (It may be permissible to argue from *un*controversial modal claims, as we might want to explain their truth in terms of more fundamental modal realities.) That analogy puts the epistemic weight on attempts to explain actuality—not extraordinary possibilities. TEM agrees. I say more about all this in Sect. 3.4.

problematic accident), there is still a difficulty about the *scope* of the mechanisms that evolutionary forces are likely to select.

What's the scope issue? The initial evolutionary argument gives some reason to think that these mechanisms are reliable when it comes to matters of survival. We still need to consider, however, whether they're reliable when it comes to matters removed from the business of everyday life. Are they, for instance, likely to produce true beliefs about highly theoretical issues? It's one thing to suggest that evolutionary forces are likely to produce a mechanism that tracks the truth about mundane matters; it's quite another to suggest that they'll produce a mechanism that rarely veers into falsehood when faced with questions outside its original purview. After all, a generally unreliable system will be reliable if it's corrected often enough, as the human cognitive system is. Where matters of survival are concerned, we receive an enormous amount of input from the world; we are constantly assaulted with information relevant to our continued existence. Hence, evolutionary forces can afford to produce a cognitive system that cuts corners, since they can bank on the world to take up the slack. But where the world offers less in the way of correction—and why shouldn't interesting modal matters be a prime example?—we have much less reason for optimism about the reliability of our faculties.[12]

Finally, if we are indeed reliable about some modal matters, our reliability is probably restricted to assessing whether a given state of affairs is physically possible. Our ancestors needed to be able to assess risks and opportunities, and our modal notions allow us to do so. If p is impossible, then we can ignore it in our deliberations; if it isn't, then we need to factor it in. And, depending on p's valence for an agent at t, p qualifies as either a risk or an opportunity for that agent at t. But it's not entirely clear why our ancestors would need to be able to assess yet stronger modalities; there is no obvious adaptive advantage to getting things right in all worlds, since, *ex hypothesi*, there were no physically impossible selection pressures. (For more on this line of reasoning, see Nozick (2003) and Nichols (2006).)

The upshot is this: non-theists are short on reasons to dismiss the possibility that we're overlooking conflicting modal claims when we try to evaluate extraordinary modal matters. So, just as the analogy with *Clue* suggests, they ought to suspend judgment where they can't draw on theories for guidance.

1.5 Looking Ahead

The game analogy suggests that realists about modality should expect to endorse something like TEM. Of course, various other considerations could point in different directions. So, we need to get a better handle on how TEM works. In the next chapter, I set out the position in detail.

---

[12]For more objections to this kind of evolutionary argument, see Stich (1990) and Plantinga's contributions to Plantinga and Tooley (2008).

Chapter 2
TEM's Details

Abstract According to TEM, a person is justified in believing an interesting modal claim, *p*, if and only if (a) she is justified in believing a theory according to which *p* is true, (b) she believes *p* on the basis of that theory, and (c) she has no defeaters for her belief that *p*. Accordingly, we need a story about how we come to justifiably believe theories, a story about theories on which they have modal content, and a story about what it is to believe a claim on the basis of a theory. This chapter provides the second and third story, and explains why I can leave the problem of theory confirmation for others.

According to TEM, a person is justified in believing an interesting modal claim, *p*, if and only if (a) she is justified in believing a theory according to which *p* is true, (b) she believes *p* on the basis of that theory, and (c) she has no defeaters for her belief that *p*. So, any version of TEM will have three main components. The first two are related to Condition (a). First, we need a story about how we come to justifiably believe theories—i.e., an account of theory confirmation. Second, we need a story about theories on which they have modal content; it needs to be case that some modal claims are true according to theories. The third component relates to Condition (b): we need a story about what it is to believe a claim on the basis of a theory—i.e., how our justification transfers from theories to specific modal claims. I'll discuss Condition (c) when considering some objections in the next chapter.

TEM's first component—a story about how we can justifiably believe (some) theories—is obviously important. Among other things, your account of theory confirmation will have some influence on whether TEM supports modal skepticism or challenges it. An account that sets a low bar for theory confirmation will make TEM more optimistic about the scope of our modal knowledge; an account that sets a high bar will be more pessimistic. And for that reason, your account of theory confirmation will affect how conservative TEM is—i.e., the degree to which it's in tension with our considered judgments about those modal beliefs that are and aren't justified. However, I don't need to say much about theory confirmation to get my project off the ground. TEM builds the epistemology of modality on the back of your epistemology for theories. If you're a scientific realist—or a naturalistically-minded metaphysician—then you need to explain how we can justifiably believe some of

B. Fischer, *Modal Justification via Theories*, Synthese Library 380,
DOI 10.1007/978-3-319-49127-1_2

our best theories. So, you have a vested interest in providing the account of theory confirmation on which TEM depends. My goal is to show that, whatever your story about theory confirmation, you have an attractive, ready-made modal epistemology in TEM.

I should make one more point. Recall one of the assumptions I made at the outset: namely, the truth of scientific realism. It follows from this that there's an adequate account of theory confirmation for scientific realists to espouse, and so TEM can deliver all sorts of technologically- and scientifically-interesting modal claims. I don't assume, however, that naturalistically-minded metaphysicians have a plausible account of theory confirmation. This isn't a backhanded way of denying that they have a plausible account: I don't know how to argue for that conclusion, and in any case, I don't know whether I believe it. Here, I'll be content with arguing at a number of junctures that it's hard to justify some of the extraordinary modal claims that have mattered to philosophers. This isn't the same as arguing that there is no good account of theory confirmation for naturalistically-minded metaphysicians, but it should be enough to defend the sort of epistemic caution that comes along with TEM. I'll say much more about all this in Chap. 5.

2.1 Theories and Their Modal Content

Enough about theory confirmation. Let's turn now to TEM's second component—the story about theories on which they have modal content—which allows us to see how TEM works. This part of the proposal is inspired by a passage from van Fraassen (1980):

> Guided by the scientific theories we accept, we freely use modal locutions in our language. Some are easily explicated: if I say that it is impossible to observe a muon directly, or to melt gold at room temperature, this is because no counterpart to such events can be found in any model of the science I accept. But our language is much subtler and richer than that; its modal locutions reflect the fact that in the models of our theories we see structures that correspond to alternative courses of events, not all of which could be jointly actualized.
>
> On the view of acceptance of theories which I have advocated under the name of constructive empiricism, it does not require belief that all significant aspects of the models have corresponding counterparts in reality. This applies to many aspects discussed by philosophers of science: space-time, elementary particles, fields, and, finally, alternative possible states and courses of events. The locus of possibility is the model, not a reality behind the phenomena (201–202).

Van Fraassen is drawing on both the semantic view of theories and his constructive empiricism here. Let's set aside the latter for the moment. The characteristic feature of the semantic view is that it takes a theory to be a family of models. This claim is what distinguishes it from the syntactic view of theories. On the syntactic view, a theory is a set of statements—ideally, ones that are expressible in first-order logic. These statements are the theory's laws, and they are interpreted by "bridge principles" or "correspondence rules" that link the theory's theoretical terms (i.e., the predicates) to observable conditions. Unfortunately, this requires a distinction between the theoretical language and the observation language, which is notoriously difficult to draw.[1] Moreover, the commitment to bridge principles or correspondence rules is based on the assumption that it's possible to translate sentences from the theoretical language into the observation language. This too is no small hurdle. (Carnap's *Aufbau* appears to be the best attempt, and the consensus seems to be that it doesn't succeed.) To make the job easier, some allowed that there could be "partial" interpretations of the theoretical language. But now there is reason to worry about the uninterpreted terms. Are they meaningless? If so, then scientists are simply making noise when they utter them, which is a hard pill to swallow. However, if the uninterpreted terms *are* meaningful, whence comes their meaning?

Admittedly, it's possible to formulate the syntactic view in a way that rejects the theoretical / observation term distinction, and so does not involve correspondence rules: you can interpret the theory's statements so that their terms refer to—and their quantifiers range over—objects and properties "out there" in the world. This eliminates the problem to which the correspondence rules lead. However, once you have the interpretation of the theory—the set of structures or models—it becomes unclear why it's important that the theory be axiomatized in one way rather than another. Wouldn't you have the same theory regardless of how you defined the set of models? Why is any particular linguistic formulation essential to it? Isn't the real question about the relationship between the models and the target system? These questions lead straight to the semantic view.[2]

Again, the characteristic feature of the semantic view is that it takes a theory to be a family of models. As Giere (1979) and van Fraassen (1980) develop it, a theory is composed of a theoretical definition and a theoretical hypothesis. The theoretical definition specifies a set of models. The theoretical hypothesis describes the purported relationship between those models and the phenomena in question.[3] If we think of that phenomena as a system, then we can say that each model bears the relevant relationship to a state of that system. As a set, the models represent all its possible states.

We can get a better feel for the semantic view by considering a toy example that van Fraassen employs. The example involves a theory of shadow in its mass-noun sense—i.e., "There is a lot of shadow in that picture"—rather than its count-noun

[1]On this point, see "What Theories Are Not" in Putnam (1979).

[2]For arguments in favor of the semantic view, see Lloyd (1994), Suppe (1977, 1989), Suppes (1993), Thompson (1989), van Fraassen (1989).

[3]The theoretical definition stands to the models roughly as sentences stand to propositions. Many sentences can express a single proposition, and it's the proposition—i.e., what's said—that's of interest. Likewise, the proposed theoretical definition is the standard way to define the set of models, but there are other theoretical definitions that would do the same work, and none has any privileged status: it's the models that matter. For this reason, it's inaccurate to say that the theoretical definition expresses the theory's *laws*. Of course, you *could* take some subset of those claims that hold true in every model of the theory to be the theory's laws, given some account of what laws are. But an adherent of the semantic view need not appeal to laws at any juncture; they need not figure into her understanding of theories, nor of their modal content.

sense—i.e., "The shadows danced among the trees" (1989, 217–218). The first step is to generate a theoretical definition. According to this theory of shadow, for any physical object x:

1. If x casts any shadow, then some light is falling directly on x.
2. x cannot cast shadow through an opaque object.
3. All shadow is shadow of something.

This definition specifies a set of models—"**S**." We don't have an explicit theoretical hypothesis—"**H**"—but since this is supposed to be a theory of shadow, we can safely assume that there shouldn't be any examples of shadow that aren't relevantly similar to a member of **S**. More formally:

H For any shadow-involving phenomenon s, there is a member of **S** to which s is isomorphic.

According to the semantic view, the theory of shadow is **S+H**.

Some modal claims are true relative to **S+H**. To see this, we can define a "*p*-model" as a model that represents the target system as being such that p is true of it, and a "representing model" as a model that's *supposed* to represent a possible state of the target system. (Not all models are representing models. If you realize that your theory idealizes the target system, you'll build that into the theoretical hypothesis, which means that the theoretical hypothesis might say that some models are just artifacts of the idealization, and so don't represent possible states of the system.[4]) With these notions in hand, we can define two more: "theoretical possibilities" (P_T) and "theoretical necessities" (N_T)—namely, the modal claims that are true according to the theory.

(P_T) p is possible according to a theory if and only if **S** includes a representing model that's a p-model.

(N_T) p is necessary according to a theory if and only if every representing model in **S** is a p-model.[5]

---

[4]There are interesting wrinkles associated with the members of **S** that *vacuously* satisfy the theoretical definition. There are a few ways to think about such cases. First, you might take it to be implicit in the theoretical hypothesis that we aren't supposed to take a stand on those models that vacuously satisfy the theoretical definition—i.e., we should neither affirm nor deny that they represent possible states of the target system. Second, you might delimit the relevant models by appealing to our explanatory aims: those models are relevant such that, without them, the theory wouldn't explain what it purports to explain. Relatedly, you might delimit the relevant models by appealing to our epistemic circumstances: those models are relevant such that, without them, we wouldn't be justified in believing the theory—the rest aren't.

[5]We can also define "theoretical counterfactuals" (C_T): if p were the case, then q would be the case according to a theory if and only if (a) **S** includes a model that is supposed to represent actuality (the "@-model"), (b) **S** includes representing p-models and representing q-models, and (c) the representing p-model most like the @-model is a representing q-model. This definition is based on Stalnaker's semantics for counterfactuals, but it's obvious that you could use Lewis's; the only difference is that Stalnaker makes the simplifying assumption that there will be a unique closest world.

Here, for example, are some theoretical possibilities and theoretical necessities for the theory of shadow:

- it is possible that a physical object cast a shadow through a non-opaque object, and
- it is impossible for a physical object to cast a shadow unless light is falling directly on it.

Of course, we might not like these implications of the theory—perhaps, for example, we think it's possible for an object to cast a shadow without light falling directly on it. But that's a reason to give up the theory, not to reject the idea that certain claims are possible or impossible according to the theory.

There are, of course, those who will want to avoid saying that this modal content represents modal *reality*: indeed, van Fraassen's constructive empiricism is designed precisely to avoid such commitments. He thinks that the goal of science is to construct theories that are empirically adequate, not to discover which theories are true. And crucially, you don't need to believe that a theory is true to accept it: you only need to maintain that it's empirically adequate. Hence, you don't have to believe that the possibilities it represents are genuine possibilities.

However, we should note that many explanations—and perhaps all—involve some sort of dependence relation: the short in the circuit caused the fire; the pressure in that container is thus and so thanks to the temperature and volume of the gas inside it; the short, sturdy beaks of the large ground finch evolved to help them crack nuts; etc. And dependence relations are modal relations: A depends on B only if, if B weren't the case, A wouldn't be the case (ignoring the ever-important *ceteris paribus* clause). Once we start articulating these modal relations, we have the resources to make claims that go beyond actuality. When we maintain that the short in the wire caused the fire, the explanation assumes that it's possible that the fire never happened: it's taken for granted in the counterfactual that the explanation underwrites: namely, that had there been no short, the building wouldn't have gone up in flames. And if we're justified in believing the explanation, we're justified in believing this assumption. Indeed, it isn't clear that you'd count as believing the explanation if you rejected the claim that the fire might never have happened. If it isn't the case that the fire might never have happened, then why do we need an *explanation* for its happening? If it was inevitable, no explanation seems necessary.

It's interesting to consider why we value explanations, so construed. One answer is that explanations simplify and order our model of the world. Kim (1994, 69) makes this point nicely:

> Dependence is asymmetric and transitive, and can generate relational structures of dependent events, states, and properties. The ontological contribution of dependence relations lies exactly in this fact: they reduce the number of independent events, states, facts, and properties we need to recognize. And that is precisely the unifying and simplifying power of dependence relations. Unity and structure go hand in hand; dependence enhances unity by generating structure.

Another answer is that we value *understanding*, which is what explanations provide. Consider, for example, Laplace's demon—a being who, given his exhaustive descriptive knowledge of the facts at *t*, can predict and retrodict every state of the world at every other time from the laws of classical mechanics. This is no small feat, but we might still reject the demon as a model of epistemic success. While the demon's computational abilities are impressive, he doesn't need to see how events are related in order to derive his results: he's just engaged in a deductive exercise. He can be as accurate without knowing what causes what (or what grounds what, etc.). This suggests that the demon lacks what explanations provide: namely, understanding through dependence relations.[6] But however we defend our interest in explanations, it remains that many of us are indeed interested in them.

In any case, since explanations provide modal information, to go van Fraassen's route is to give up on explaining lots of important phenomena, settling for mere prediction instead. Van Fraassen won't postulate quarks because we can *predict* perfectly well without them, even if we can't *explain* without them. This is a non-starter for those who care about explanation, as realists certainly do.

So if we're realists, then we can use the semantic view as the basis of a modal epistemology. We will, of course, be obliged to defend our commitment to explanation and our preferred account of theory selection—whether inference to the best explanation, or conditionalization, or what have you—but no realist thought otherwise. Once we discharge this obligation, the semantic view does the remaining work. Our account of theory selection explains how we can be justified in believing that certain theories are true; the semantic view explains both how our theorizing generates modal commitments, as well as which commitments it generates.

Let's suppose, then, that when you believe a theory to be true, you believe that a class of models represents a system. Now we can be more precise about TEM. A theory commits us to a modal claim if *p* is possible or necessary according to that theory. And if a theory commits you to a modal claim, then your reasons to believe the theory are reasons to believe the modal claim. So, for example, if you justifiably believe a theory according to which *p* is possible (in the sense of P_T, above), then—absent any defeaters and assuming that you believe that *p* is possible on this basis—you are justified in believing that *p* is possible. This leads us to the following conditionals:

> [POSS] Where *p* is any interesting modal claim, you justifiably believe that *p* is possible if and only if (a) you justifiably believe a theory according to which *p* is possible, (b) you believe that *p* is possible on the basis of this theory, and (c) you have no defeaters for the belief that *p*.

[6] If you doubt that the demon lacks understanding, we can get a conclusion that's just as good for present purposes via a slightly different argument. Suppose, again, that the demon can perfectly predict the evolution of a system, and that the demon is kind enough to report his predictions on demand. (He's always truthful: if the demon says that an object will be at a location at *t*, the object is at the location at *t*.) With the demon available to you, you too can predict and retrodict every state of the world at every other time. But clearly you aren't a model of epistemic success in this scenario: you don't need to understand anything at all to have this power. Explanation is a good that goes beyond mere predictive (and retrodictive) accuracy. For further problems with the view that explanation is prediction and retrodiction, see Salmon (1989).

> [NEC] Where *p* is any interesting modal claim, you justifiably believe that *p* is necessary if and only if (a) you justifiably believe a theory according to which *p* is necessary, (b) you believe that *p* is necessary on the basis of this theory, and (c) you have no defeaters for the belief that *p*.

TEM, then, is the conjunction of [POSS] and [NEC], which we can state more succinctly if we gloss over the differences between the two conjuncts:

> [TEM] Where *p* is any interesting modal claim, you justifiably believe that *p* if and only if (a) you justifiably believe a theory according to which *p* is true, (b) you believe that *p* on the basis of this theory, and (c) you have no defeaters for the belief that *p*.

2.2 Interlude: The Syntactic View Revisited

I've developed TEM using the semantic view of theories, but I should acknowledge that it isn't necessary: you could get a version of TEM without it. Admittedly, you will thereby sacrifice the benefits associated with the expressive power of the semantic view, but you'll gain simplicity when discussing metaphysical theories, where the expressive power of the semantic view generally isn't that important.

To see what I mean, consider the following passage from Kripke (1980, 113):

> How could a person originating from different parents, from a totally different sperm and egg, be *this very woman?* One can imagine, *given* the woman, that various things in her life could have changed: that she should have become a pauper, that her royal blood should have been unknown, and so on. One is given, let's say, a previous history of the world up to a certain time, and from that time it diverges considerably from the actual course. This seems to be possible. And so it's possible that even though she were born of these parents she never became queen. Even though she were born of these parents, like Mark Twain's character she was switched off with another girl. But what is harder to imagine is her being born of different parents. It seems to me that anything coming from a different origin would not be this object.

This passage ends with the necessity of origin thesis, which is a metaphysical theory. According to that thesis, certain features of a thing's origin are essential to it. Had those features not been present, we'd have a different individual on our hands. And in principle, we could express this view using the semantic approach to theories: we could try to distinguish a theoretical definition and theoretical hypothesis, let the former define a set of models, and so on. But this might feel overly complicated. The necessity of origin thesis is just that—a thesis. We don't need the semantic view's machinery to understand it.

The reason for this is that metaphysical theories are much more like van Fraassen's toy-theory of shadow than they are like a theory of predator-prey dynamics. This isn't a side-swipe at metaphysical theories, but an observation about their theoretical hypotheses—namely, that they're all the same. I suggested that the theoretical hypothesis for the theory of shadow was something like:

H For any shadow-involving phenomenon *s*, there is a member of **S** to which *s* is isomorphic.

Isomorphism is probably too loose a relation for metaphysical purposes. So if we wanted to express the necessity of origin thesis using the semantic view, we might refine H as follows (recalling that the thesis defines a set of models):

$\mathbf{H_{origins}}$ For all *x*, *x* is a member of **S**.

The upshot of $\mathbf{H_{origins}}$ is just that everything either trivially or non-trivially satisfies the theoretical definition of the necessity of origin thesis—or, more colloquially, that the necessity of origin thesis is true.[7] This simplification is tolerable because metaphysicians don't take themselves to be offering idealized or partial theories; they also rarely describe the world using mathematical models. Instead, they take themselves to be offering, in natural languages, literally-true descriptions of how things are. (At this point, we're essentially dealing with a liberalized version of the syntactic view of theories relative to its positivist origins.)

This is a feature, not a bug. Suppose we want to explain how someone might come to justifiably believe that, say, nothing can travel faster than the speed of light, or that (contra van Inwagen) there could be naturally purple cows, or what have you. Then, we need an account that's sensitive to the ways that we formulate physical theories. I've framed TEM in terms of the semantic view of theories because there's more to modal epistemology than the epistemology of extraordinary modal claims. Interesting modal claims abound, and we need an account of how we know them when we *do* know them. Granted, then, the machinery I'm developing doesn't really matter for many of the claims that concern philosophers. What *does* matter is that whatever our account of theories, we be able to extract modal information from them. We can get that information from the semantic view via the method sketched above. But of course, if the theory already involves modal operators (as we find, e.g., in the necessity of origin thesis) then we don't need to rely on anything so complex: we can simply deduce the theory's modal consequences. Or, to make the same point in a different way, TEM's first condition requires that you believe a theory according to which *p* is true, but it isn't important *how* the theory specifies that a proposition is true. What counts is that it does.

2.3 The Basing Condition

To be justified in believing an interesting modal claim, TEM requires that you be justified in believing a theory according to which it's true. To make this work, we need a story about how we come to justifiably believe theories (an obligation I've pushed off onto others) and a story about theories on which they have modal content (which the semantic view provides, though other accounts are available). But TEM

[7]If we wanted to bracket the trivial cases, we could restrict $\mathbf{H_{origins}}$ to things that have origins, but nothing turns on this here.

doesn't merely say that you need to be justified in believing the relevant theory—you also believe that the interesting modal claim on the basis of that theory. Can we spell out this condition in a plausible way?

I think so. Even better, we can even use this clause to explain the epistemic credentials of conceiving, imagining, and intuiting. In short, we can say that these mental activities are not the sources of our justification about modal matters, but rather are the means by which we identify the models of our theories. For example, if you have the intuition that *p* is possible, we can say that this is an especially quick judgment to the effect that there is a *p*-model specified by one of the theories you justifiably believe. (Or, if you're committed to the syntactic view, an especially quick judgment to the effect that *p* follows from a theory that you justifiably believe.) Conceiving and imagining offer the same derivative justification, although they do so more slowly, more deliberately, and by way of more or less sensuous imagery. Hence, believing *p* on the basis of a theory doesn't require conscious reflection on a theory. We can maintain that conceiving, imagining, and intuiting play important roles in the epistemology of modality without allowing that they're the basic source of our justification. If TEM is correct, they are downstream from that source—namely, the arguments that we give for our best theories.

This sort of view is not original. Many naturalistically-minded philosophers have argued that intuitions confer justification only because they are appropriately related to something else we know. Consider, for example, the view that David Lewis defends with respect to the imagination:

> We get enough of a link between imagination and possibility, but not too much, if we regard imaginative experiments as a way of reasoning informally from the principle of recombination. To imagine a unicorn and infer its possibility is to reason that a unicorn is possible because a horse and a horn, which are possible because actual, might be juxtaposed in the imagined way (1986, 90).

For Lewis, we certainly need to defend the principle of recombination; it isn't a Moorean fact. But after we've defended it, we can offer sympathetic reconstructions of how others modalize. Likewise, I think that people can be justified in believing modal claims if they get to them by drawing on the theories that they justifiably believe whether or not they draw on those theories explicitly.[8]

Is it plausible that we draw on theories even implicitly, or does TEM condemn much of our modalizing? There's cause to think the former.

Consider the approach to cognition known as "the theory of mental models."[9] On this view, we reason by way of what are, essentially, iconic versions of the classes of models that the semantic view employs. Here is how P. N. Johnson-Laird summarizes one of the theory's key tenets:

> A principle of the modern theory of mental models is that a model has the same structure as the situation it represents. Like an architect's model, or a molecular biologist's model, the parts of the model and their structural relations correspond to those of what it represents.

---

[8]For a related view, see DePaul and Ramsey (1998).

[9]See Gentner and Stevens (1983) and Johnson-Laird (2004) for helpful overviews.

> Like these physical models, a mental model is also partial because it represents only certain aspects of the situation. There is accordingly a many-to-one mapping from possibilities in the world to their mental model (2004, 181).

In the above passage, Johnson-Laird may mean only that there is a one-many relationship between the mental model and the target systems. (The mechanic does not have one mental model for each engine he repairs, but one mental model that applies to a number of engines.) However, it's plausible that he intends more: namely, that on the side of the many, some of the relata are non-actual states of affairs—ways that the target systems could be. This suggests a natural relationship between theories and mental models: mental models probably aren't *identical* to theories, at least as the semantic view interprets them, but they may still be best *represented by* theories. We can suppose that a mental model specifies (more or less exactly) a set of models by way of the structural similarities that Johnson-Laird mentions. Some of those models represent actual states of affairs, but many will represent non-actual states of affairs. And we can interpret our attitudes toward our mental models as roughly equivalent to theoretical hypotheses: if we endorse every implication of a given mental model, then we suppose that there is an exact correspondence between our mental model and the target system; if we withhold judgment in some cases, then we suppose that our mental model is imprecise in some respect or other. The upshot is that the theory of mental models appears to sit comfortably with the semantic view of theories; it seems to provide an attractive account of what's in the head when you believe a theory—namely, a mental structure that corresponds to a class of models.

Here's a toy example. Suppose that we're trying to determine how many bottles of beer the fridge will hold. According to the theory of mental models, what we have in mind is a structure that in some way resembles the fridge and bottles. Some features of the structure are invariant, but others are, essentially, moving parts—i.e., features of the structure that can be altered, or that are removable, allowing new features to be introduced. Presumably, the walls of the fridge correspond to the invariant aspects of the mental model, while the moving parts—the shelves, the drawers, the bottles—correspond to the variable components. This structure represents, more or less precisely and more or less accurately, all the ways that beer can be arranged in the fridge. If the mental model theory is correct, then we reach a conclusion about how best to pack the fridge by manipulating this mental structure.[10]

Let's assume that the mental model theory is true. Then, we can say that you believe *p* on the basis of a theory if you believe *p* as a result of manipulating a mental structure that corresponds to the appropriate class of models—a structure that you possess, we'll presume, in virtue of having developed or learned the relevant theory.[11] As I've already suggested, this is a promising account of the

[10]For a related view, see Nichols (2006).

[11]Of course, if you aren't sufficiently familiar with the theory in question, you won't be able to draw on it in the way I'm suggesting. But if you aren't sufficiently familiar with the theory in question,

cognitive machinery behind conceiving, imagining, and intuiting—the standard routes to justified belief about modal matters—about which we can now be more precise. So, for example, we might say that to conceive or imagine that *p* is to manipulate a mental structure so that it corresponds to a *p*-model. Depending on a variety of factors, this may or may not involve a number of mental images; hence the distinction between conceiving and imagining. Intuiting that *p*, on the other hand, might be either (a) a species of conceiving or imagining—perhaps just the limit case where speed is concerned—or (b) to come to believe *p* based on some heuristic device, itself just a rough and ready approximation of a more complex structure.

Of course, much more needs to be said about the theory of mental models, its purported fit with the semantic view, and our modalizing. Still, this seems like a promising way to forge connections between TEM, cognitive psychology, and the mental operations that have long held pride of place in the epistemology of modality.

2.4 Can TEM Be Extended?

Earlier, I drew attention to Vaidya's (2015) distinction between uniform and non-uniform modal epistemologies—the former posit a single source of our justification concerning modal matters; the latter posit multiple sources. TEM is a non-uniform account: it only purports to explain interesting modal justification. But could it be extended? Could it be made into a uniform modal epistemology for all modal claims? As I finish laying out TEM's details, I want to explain both how that might be done and why I'm not doing it.

The obvious problem with extending TEM is that it would result in a fairly wide-ranging skepticism about those modal claims of which we're most confident. Imagine a version of TEM according to which, where *p* is *any modal claim whatever*, you justifiably believe that *p* only if you justifiably believe a theory that says that *p*. What theory says that I could have had coffee with breakfast this morning, or that my desk could be a foot to the left, or that my shirt could be striped instead of plaid? If we're justified in believing any modal claims, we're justified in believing these ones. TEM, however, seems ill-equipped to explain why.

We might take a first step toward a solution by taking a more generous view of what counts as a theory. Indeed, we might say that *all* our beliefs somehow "amount to" a theory. (Quine takes a position like this.[12]) Of course, it matters what "amount to" amounts to in this context. Is it the case that our beliefs *are* theories (i.e., each belief is a mini-theory)? Alternately, do our beliefs jointly *constitute* a theory—or a family of theories? Either way, perhaps we believe the theory in virtue of having the

TEM says that you can't use to secure justified beliefs about the modal claims it sanctions, so this is hardly a problem.

[12]For similar perspectives, see Lewis (1970), Churchland (1986), Churchland (1989), and Baker (2007).

beliefs. On the other hand, maybe "amount to" ought to be construed more broadly. Are our beliefs merely *an expression* of a theory—and hence are distinct from it? Or are our beliefs simply *evidence* of a theory—perhaps the one that we would accept if we were presented with an explicit version of it? (These interpretations seem to fit the way that descriptive metaphysicians think about our beliefs, since they take our theory of the world to be something about which we need to be informed; see, e.g., Strawson (1959)). On the broader reading, it's less clear whether we even *believe* the theory in question, much less whether we're *justified* in believing it. Hence, it's not obvious that these interpretations are of use to TEM, at least in its present form. However, if pressed to opt for one of these interpretations, then it may be possible to modify TEM accordingly—maintaining, say, that you justifiably believe *p* only if you stand in the appropriate relationship to a theory that says that *p*, where the precise character of that relationship depends on the interpretation that wins the day. However these details are sorted out, let's say that *folk theory* is the theory that is alleged to stand in the relevant relation to all those beliefs that we do not obviously hold on the basis of some scientific or metaphysical theory[13]—such as the belief that my shirt could be striped instead of plaid—and let's say that *the folk theory hypothesis* is the suggestion that those beliefs *do* stand in the relevant relation to folk theory.

To be clear, some think of the folk theory hypothesis as an *empirical* hypothesis about parallels between the acquisition, structure, or function of our ordinary beliefs and the acquisition, structure, or function of our scientific theories; see, e.g., Gopnik and Meltzoff (1997). This may well be right, but it isn't the only strategy for the TEM-extender to pursue. Instead, the TEM-extender could understand it as an epistemological claim: however we acquire our ordinary beliefs, they are *justified* (or at least justifiable) in the same way as our scientific and metaphysical beliefs. On this view, there is no difference in kind between the arguments that we offer for scientific and metaphysical theories and the arguments that we (could) offer for folk theory. (An exemplar might be the way that Jonathan Vogel (1990) argues that the Real World Hypothesis is a better explanation of our experience than is any of the scenarios proposed by the Cartesian skeptic.) At the same time, the TEM-extender could understand it as a claim about how we should think about the content of folk theory: folk theory can be represented just as we represent scientific theories, which is to say that we can treat folk theory as a set of models. The epistemological aspect allows someone to say that folk theory can also be justified however our scientific theories are justified; the bit about representing folk theory lets an extended version of TEM handle even our uninteresting modal knowledge.

---

[13]Or, if you prefer, let's say that folk theory is that part of our global theory that stands in the relevant relation to all those beliefs that we do not obviously hold on the basis of a scientific or metaphysical theory. Alternately, let's say that our folk theories are those theories that stand in the relevant relation to various sets of beliefs not obviously held on the basis of a scientific or metaphysical theory. I'll ignore these variants in what follows, since I don't think that much turns on how we individuate theories.

The next step is to argue that folk theory has something to say about the examples with which we began. After all, surely there are matters about which folk theory is silent, or ambiguous, or even flatly inconsistent. If the theory is silent with respect to (say) p's possibility, then TEM says that we ought to abstain from belief about it; if the theory is vague, then we will need to make it more precise before we take a stand; if the theory is inconsistent, then we will need to repair it before we believe one way or the other. Here are some tentative proposals about the examples with which we began:

- *I could have had coffee with breakfast this morning.* We appeal to beliefs and desires to explain our actions, and behind these explanations is the assumption that different belief/desire combinations would have led to different actions; the belief/desire pairs explain because of the way that actions (supposedly) depend counterfactually on them. It's plausible, therefore, that folk theory includes some folk psychology, part of which is a primitive libertarian theory of the will. So, my guess is that folk theory has models in which (a) I wanted coffee with breakfast this morning and (b) I successfully acted on that desire. Hence, folk theory has models in which I have coffee with breakfast this morning.
- *My desk could be a foot to the left.* We spend our lives navigating a world full of medium-sized objects. Accordingly, we have good reason to develop a working theory of what they can do and what can be done to them. This theory will be no good if it says that, at a time, no object can occupy a location other than the one it in fact occupies. Either the theory says that I have the ability to move my desk, or it doesn't. If I have this ability, surely I can exercise it at any given time (all other things being equal). And if I don't have this ability, how do we explain the cases in which I successfully slide my desk to a different spot? So, it's quite likely that folk theory has models in which my desk is a foot to the left.
- *My shirt could be striped instead of plaid.* This sentence is ambiguous. On perhaps the simplest reading, the claim is that I could be wearing a different shirt—namely, one with stripes. This makes it much like the claim that I could have had coffee with breakfast this morning; presumably, our theory of the will allows that I have these sorts of options available to me. Read another way, the sentence says that the plaid pattern could be replaced with stripes. This interpretation makes it akin to the assertion that my desk could be a foot to the left: we know that enough bleach will remove the existing pattern; fabric markers will do the rest. But there is yet a third reading: understood this way, the allegation is that my shirt could *always* have been striped rather than plaid. This is perhaps the most interesting case, since it's not obvious what folk theory says, if it says anything at all. This is no objection, though, because silence is probably the right result. Could my shirt have had a different pattern? Could it have been made from different cloth? Could some of its threads have been of a different color? Could it have been made elsewhere by different people? Could it have been made by a different company? It isn't clear how to answer these questions. Or, more accurately, it isn't clear how to answer these questions without invoking a contentious metaphysical thesis, such as the necessity of origin, or mereological

essentialism, or what have you. But the present view lends itself to an explanation of why this is so: we are rarely confronted with situations in which we need to settle questions like these, and so folk theory is not well-equipped to handle them. Of course, we could *extend* folk theory so that it *does* answer them—and alongside precisifying and repairing folk theory, that is just the sort of work to which metaphysicians devote themselves. At any rate, there are some readings of *My shirt could be striped instead of plaid* that folk theory seems to affirm fairly unambiguously, and there is at least one where matters are much less clear, but the clarity isn't an objection to the project of extending TEM.

Of course, philosophers are free to reject any of the theories that I just used to get these results. No doubt many will want to want to reject the libertarian component of folk psychology, for example. The point is just that someone who wants to extend TEM can argue in the way I've suggested, and that it isn't obviously unreasonable. Perhaps we have no better way to understand and predict macroscopic events than by appealing to ships and shoes and sealing wax, and so have reason to believe that the relevant theory of ordinary objects is true. Likewise for our theory of the will. And, if we are justified in believing these theories, then TEM allows us to use them to justify various mundane modal beliefs.

So what's the problem? Why not extend TEM? The short answer to this question is that the extended version of TEM stands or falls on an empirical issue that's beyond the scope of this project. The longer answer is as follows.

TEM is a biconditional, and I take it that an extended version of TEM would preserve that structure. TEM, of course, restricts itself to interesting modal claims, whereas the extended version wouldn't. However, if we're going to make an extended version of TEM plausible, then it isn't enough to argue that we could be justified in believing uninteresting modal claims on the basis of theory—you'd have to argue that we *are* justified in believing them on this basis. This is because we're assuming that we are in fact justified in believing them. So, if we didn't come by this justification as a result of relying on theories, then an extended version of TEM couldn't be a biconditional: it would need to say that while you're justified in believing a particular uninteresting modal claim if the relevant conditions are met, it isn't the case that you're justified in believing it only if they're met. And since it would allow other sources of justification, it wouldn't be a unified account, and we wouldn't get the simplicity and elegance that makes a unified account especially attractive. So, to make it worth it to extend TEM, you need to argue that we are, in fact, justified in believing ordinary modal claims based on theories. How optimistic should we be about making that argument?

Not very. To see the problem, we should note that the phenomenology of our ordinary modal judgments doesn't directly support an extended version of TEM. When asked whether the table could be to the left of where it actually is, it's clear that I don't explicitly consult a theory of medium-sized dry goods. If I think about the answer at all—which I might not, as I might all-but-automatically answer in the affirmative—I'm probably imagining the table's being in a different location.

On the face of it, then, either intuition or the sensuous imagination does the work. And given that phenomenological considerations point toward intuition and / or imagining—which may be poor guides to modal matters generally, but perfectly fine guides concerning various uninteresting modal matters—someone who wants to argue for an extended version of TEM can't appeal to a presumption in favor of theories doing the work. He's going to have to argue for that in some other way.

To be clear, none of this is to suggest that the phenomenology of our ordinary modal judgments is incompatible with an extended version of TEM, for reasons I've already outlined. If we adopt the theory of mental models, we have a way of explaining the mental operations behind our modal judgments in a way that fits with the idea that theories are behind them. And if we take on the folk theory hypothesis, we'll have the theories we need to underwrite ordinary modal claims. The point is just that an extended version of TEM gets no support from phenomenological considerations.

In any case, how else might someone argue that theories are, in fact, doing the work? Presumably, we'll need empirical evidence: namely, data supporting the theory of mental models, as well as whatever ancillary hypotheses we need to conclude that theories are, ultimately, explaining how most of reach conclusions that we judge, on independent grounds, to be justified. I'm a philosopher, not a cognitive scientist, and I have no independent arguments to offer for any of the relevant empirical claims. So, though it's tempting to extend TEM, that won't be my aim here. The case for an extended version of TEM turns on questions that are beyond my expertise.

2.5 Conclusion

TEM is an account of the standard to which realists should hold interesting modal claims. We now have most of it on the table: on the one hand, a story about theories on which they have modal content; on the other, a story about how we extract that information. I haven't said much about theory confirmation, but I've explained why I don't think I need to, and I'll discuss some concerns about that stance shortly. The only thing I've ignored is TEM's no-defeater condition.

There are two types of defeaters: rebutting and undercutting. A rebutting defeater for your belief that p is a reason to believe that $\sim p$; an undercutting defeater for your belief that p is a reason to think that your belief was formed improperly—e.g., on the basis of false testimony, or sloppy reasoning, or what have you. According to TEM, then, you aren't justified in believing an interesting modal claim if it's been either rebutted or undercut. However, this general point applies to all our beliefs, and so isn't of particular interest in this context. So, I won't spend more time on it. Instead, I want to have a look at some specific defeaters that are of concern because of TEM's commitments. These will come into view as we examine objections to TEM in the next chapter.

Chapter 3
Objections and Clarifications

Abstract This chapter replies to several objections to TEM. Among them: that physical theories can't give us the kind of justification that TEM says they can, that TEM is circular, that TEM doesn't fit with a standard way of arguing in metaphysics, that TEM is far too cautious about our modal justification, that TEM can't handle conflicts between theories, and that TEM is far too permissive in terms of our modal justification.

TEM says that we're justified in believing an interesting modal claim, *p*, if and only if (a) we justifiably believe a theory according to which *p* is true, (b) we believe *p* on the basis of that theory, and (c) we have no defeaters for the belief that *p*. In this chapter, I fend off some objections to TEM's viability.

3.1 Physical Theories Don't Say Anything About What's Metaphysically Necessary

The first objection concerns the content of physical theories. The thought is this. If *p* is physically possible, then it's metaphysically possible. So, if physical theories can tell us what's physically possible, then they can tell us something about what's metaphysically possible. POSS—the part of TEM that concerns possibilities—is not controversial on this score. However, NEC is—i.e., the part of TEM that concerns necessities. In particular, someone might doubt that our physical theories have anything to say about what's metaphysically necessary.

We might offer a conciliatory response. We could, for example, introduce varieties of necessity, so that you're justified in believing that *p* is physically necessary if a physical theory says that *p* is necessary, and justified in believing that *p* is metaphysically necessary if a metaphysical theory says that *p* is necessary. (We could get roughly the same effect by introducing a restriction: though you can justifiably believe that *p* is possible based on any sort of theory, you can only believe that *p* is metaphysically necessary based on a metaphysical theory.)

However, I think we can push back. At first blush, anyway, the challenge depends on the claim that since physical regularities are contingent, we should interpret

B. Fischer, *Modal Justification via Theories*, Synthese Library 380,
DOI 10.1007/978-3-319-49127-1_3

physical theories as concerning an appropriately restricted set of possibilities. But that's just to insist that we're justified in believing a particular metaphysical theory that has implications for how we interpret all physical theories—namely, one according to which physical regularities are contingent. If that's true, then it follows that the metaphysical theory provides us with a defeater for taking any necessity sanctioned by a physical theory to be a metaphysical necessity. And, of course, if we *are* justified in believing that metaphysical theory, then we do have such a defeater, and hence NEC never said we were justified in believing the necessities in question. In short: either we justifiably believe that physical regularities are contingent, or we don't. If the former, then NEC doesn't say we're justified in believing that, for example, the speed of light couldn't have been otherwise. If the latter, then there is no objection.[1]

Someone might try to weaken the challenge: the claim isn't that we're justified in believing that physical regularities are contingent. Rather, it's that we *aren't* justified in believing that they *aren't* contingent. That is, it's an open question whether physical regularities are contingent, so we should interpret physical theories cautiously, restricting their scope.

There are a few points to make here. First, there's a sense in which even this version of the challenge smuggles in a theory for which we'd like an argument. It isn't an open question whether physical regularities are contingent unless we've *already* got a theory of modality on which the physical possibilities are a subset of the metaphysical possibilities, and the issue now is whether they're a proper subset. If you're justified in believing such a theory, then you might have reason to interpret physical theories cautiously. If you aren't, then you might regard as equally likely that the *metaphysical* possibilities are a subset of the *physical* possibilities. And if you *were* to regard this as equally likely, then surely the appeal to caution would seem unmotivated. To be clear, my point isn't that this is a plausible account of the relationship between physical and metaphysical possibility. Rather, the point is just that we shouldn't ignore how certain assumptions shape this objection to TEM, and those assumptions are in need of justification. Of course, once we argue for them, the objection to TEM disappears, as we're then justified in believing a relevant theory.

Second, even *if* you're justified in believing that the physical possibilities are a subset of the metaphysical possibilities, it isn't clear that you've got reason to interpret physical theories cautiously. Making that claim presupposes a sharp distinction between physical and metaphysical theories, which methodological naturalists reject. Moreover, it presupposes that we can reliably distinguish physical from metaphysical theories—or, more plausibly, the parts of the theory that merely make claims about the scope of the physical possibilities and those parts that make more sweeping claims about the scope of the metaphysical possibilities. It isn't clear

[1]Of course, you might have a view on which only *some* physical regularities are contingent: e.g., "All physical bodies attract one another according to the gravitational constant" is and "All samples of water are samples of H_2O" isn't. But presumably, you also have some story about why these regularities are different (e.g., the latter is underwritten by identity, whereas the former isn't), and you can use this story to guide when you restrict the interpretation of the relevant theories.

that either assumption is true. Moreover, even if someone is inclined to accept them, TEM is an attempt to provide a naturalistically-acceptable modal epistemology. So, although the discontinuity assumption may provide someone with a good reason to reject TEM, it doesn't provide any reason to think that TEM fails on its own terms.

The third point, however, is the most important. TEM is based on the idea that our best theories provide our only guide to interesting modal matters. So if we've got a theory according to which metaphysical possibility outstrips physical possibility, then we won't want to say that things must be as our physical theories say they must be. Likewise, if theories *aren't* our only guide to interesting modal matters—if TEM is mistaken, and we can come to know about physically-impossible metaphysical possibilities through unaided conceiving, or whatever—then we won't want to say that things must be as our physical theories say they must be. However, *absent* such a theory, and on the assumption that the game analogy is a good one, we have no reason to doubt that our physical theories describe the limits of what's possible—i.e., that they describe both what's possible and what's absolutely necessary. So while we could weaken TEM so that it tells us to remain agnostic about the necessities specified by some theories, that move is entirely unmotivated unless (a) you're granting TEM, and you just take yourself to have a good argument for a certain metaphysical theory, or (b) you're rejecting TEM, in which the objection begs the question. If the former, there's no disagreement. If the latter, there's no objection.

The upshot is as follows. Either you are, or are not, justified in believing a theory that gives you reason to qualify your interpretation of physical theories. If you are, then NEC doesn't say that you're justified in believing that p is metaphysically necessary because a physical theory says that p is necessary. If you aren't, then insofar as you have reason to believe that the theory is true, you have reason to think that what it says to be necessary is necessary *simpliciter*.

3.2 If TEM Is True, We Can't Justifiably Believe Physical Theories

At this point, someone might worry about whether we justifiably believe physical theories. After all, if they have things to say about what's metaphysically necessary, why think that even broadly empirical evidence for them is sufficient? I have two things to say about this objection.

First, there is no objection if we're justified in believing a metaphysical theory that leads us to restrict our interpretation of physical theories. Of course, we then face familiar puzzles about how metaphysical theories are justified. But according to TEM, physical theories don't imply anything about what's metaphysically necessary if we have antecedent reasons to restrict our interpretation of physical theories. TEM just leads us to ask the right question: do we, or do we not, have such reasons?

Second, and the above aside, it's very hard to make this sort of objection stick without resorting to a style of argument that leads to skepticism about unobservable objects. Essentially, the complaint is that we're dealing with underdetermination of theory by evidence. But scientific realists have long known this, and they still deny that we should therefore abandon our commitment to muons and gluons. Likewise, there is underdetermination of theories-having-implications-about-what's-metaphysical-necessary by evidence. So perhaps our justification is very weak, or especially fragile, or is such that we only have it in certain contexts, or whatever. Again, though, we've long known this, and any realist epistemology has to deal with it.

3.3 The Circularity Worry

So the metaphysical implications of physical theories are no barrier to their justification. However, some might think that you need to know how things could be before you can know how things are. That is, someone might insist that we need to know which theories are possibly true before we can assess which theories *are* true, which means that some modal knowledge needs to precede knowledge of actuality. If that's right, then it's backwards to try to explain modal justification in terms of our justification concerning theories, as TEM recommends.

This objection asks too much of candidate hypotheses. We can distinguish between (a) knowing that a theory is possibly true, (b) knowing that a theory is *not* possibly true, and (c) not knowing whether a theory is a possibly true. Clearly, you should not take seriously those hypotheses that you know to be impossible. However, if you don't know whether a theory is possibly true, then you don't yet know that it's off the table. It's still a live hypothesis. And, assuming that it fares well enough on the criteria for theory selection, then you have reason to think that the world is as it says, which implies that the world could be as it says. So, there is room for the order of justification that TEM recommends.

Of course, someone might say that the circularity worry isn't about the possibility of the hypotheses, but about how we assess theories. Perhaps our methods themselves require some modal information. This, it seems, is E. J. Lowe's (2008) view. He considers how we might know that Hesperus is Phosphorus—i.e., that they are, in fact, identical. He claims that *if* you know that Hesperus is Phosphorus, then you know antecedently that they "are material objects and that any such object necessarily excludes another" (26). After all, suppose not. Then, he can run the following argument:

1. Suppose you don't know that Hesperus and Phosphorus are material objects and that any such object necessarily excludes another.
2. Then, you should take it to be possible that they *don't* necessarily exclude each other—i.e., you should take it to be possible that they overlap, perfectly and always.
3. No empirical evidence could show that Hesperus and Phosphorus *don't* overlap perfectly and always.

4. So, if you don't know that Hesperus and Phosphorus are material objects and that any such object necessarily excludes another, then you can't know that Hesperus and Phosphorus are identical based on empirical evidence.

The second premise seems plausible enough, so the third is where the action is. Lowe tries to motivate it by contrasting different kinds of entities, which is supposed to reveal that certain modal commitments inform how we interpret our experiences:

> ... if ancient astronomers had believed that Hesperus and Phosphorus were not material objects of any kind but, instead, holes in the firmament through which the cosmic light shines, then they would have had no reason at all to identify Hesperus and Phosphorus even if they had been able to predict that, at a certain time, Hesperus and Phosphorus would exactly coincide. For holes are things of such a kind that they can coincide without being identical—as are, for example, shadows and spots of light. We can readily make sense of the thought that two holes in a surface should approach one another, merge for awhile, and then carry on travelling their separate ways, just as we can make sense of the thought of two shadows or spots of light merging and then separating again. Not so with two material objects of the same kind (27–28).

However, this isn't particularly convincing. I don't know the identity conditions for holes, shadows, or spots of light, but I don't see why we should think that our beliefs about them are informed, first and foremost, by antecedent modal commitments. One striking think about shadows is that they don't explode, deform, or ricochet off one another when they collide—indeed, that's why we don't generally think of shadows as *colliding* at all. What's more, it looks like shadows can merge, go their separate ways, and be recognizable as their former selves. This is evidence that they can overlap. By contrast, we have evidence for the opposite conclusion concerning most material objects: most material objects *do* explode, deform, or ricochet off one another when they collide, and they don't survive mergers and separations without incident or injury. What's more, we've got no independent reason to posit overlap: absent evidence for overlapping turtles, the simpler supposition is that what looks like one turtle *is* one turtle. So, we think that material objects exclude one another, and it looks like the evidence for this is empirical.

The upshot is this. Lowe is right that empirical evidence won't decisively *prove* that there aren't overlapping planets, but we're all fallibilists now, and we shouldn't be bothered by this. Instead, we should say that empirical evidence can give us good reason to prefer some explanations over others. But to explain is to provide modal information. When we want to explain an actual event, *b*, we say that *a* caused *b*, or that *b* "just is" *a*, or that *b* otherwise depends on *a*—all of which are modal notions that ground certain counterfactuals. So it's already part of the realist framework that we explain non-modal phenomena using modal notions, and once we've begun, we can bootstrap from there. (For naturalistically-minded metaphysicians, this means that they're rightly focused on explaining either actual phenomena or the modal phenomena that we've postulated to explain actual phenomena.) As I mentioned earlier, we do need to say something about why we care about explanation, as opposed to, say, mere prediction. But whatever we say in defense of our epistemic aims, it isn't viciously circular to draw on theories to explain modal justification.

3.4 TEM and Metaphysical Methodology

Consider the following argument:

1. I could exist without a body.
2. If I could exist without a body, then I am an immaterial thing.
3. So, I am an immaterial thing.

This argument reaches a striking metaphysical conclusion (substance dualism) based on a modal premise (namely, that I could exist without a body). And although the modal premise is controversial, many metaphysicians seem to think that this is a legitimate and important way to argue for metaphysical theses: you begin with a plausible modal claim, you point out that it has an interesting metaphysical implication, you draw the conclusion. However, if TEM is correct, then this method gets things backward. And insofar as the method seems like a good one, that's a strike against TEM.

There are two moves to make at this juncture: the first, to do some damage control; the second, to reject the idea that damage control is needed. Let's begin with damage control.

There is a way to think of TEM being compatible with the above methodology, albeit a way that will be disappointing to those who run such arguments. For the compatibility, suppose you justifiably believe a theory according to which you could exist without a body. Then, of course, you can be justified in believing the argument's first premise, and so draw the interesting conclusion (assuming that you're justified in believing the second premise). This matters because we aren't always clear about what our theories commit us to, and it can be striking to discover that our understanding of the self commits us to a substantive metaphysical conclusion. Of course, our understanding of the self does the work here, and in that sense, this methodological compatibility will be disappointing to someone who wants to avoid having to defend a theory about the self on non-modal grounds. However, it isn't clear how we could entirely circumvent our understanding of the self when trying to make arguments about the self, so perhaps the desire to do so is misguided. Either way, it's worth remembering two things.

First, TEM *can* provide a route to the interesting conclusions themselves. If you have some metaphysical commitments, then you'll want your epistemology to say that the relevant beliefs are justified. This TEM can do, assuming that you take yourself to be justified in believing a relevant metaphysical theory. If you are, then TEM says that you can be justified in believing the modal claims that it sanctions. So, *if* you think that you have a good argument for physicalism, then you can use TEM to say that you are justified in denying that there could be zombies, or that we might be disembodied. And *if* you think that you have a good argument for dualism, then you can use TEM to say that you are justified in taking the opposing stances. Crucially, though, TEM puts the weight on the arguments for metaphysical theories. If TEM is correct, then those arguments are rarely going to be modal arguments, bootstrapping aside. If the justification of your modal beliefs depends on the justification of (some of) your theories, then you will rarely be in a position to defend a theory based on modal claims. That would be to put the cart before the horse—unless, of course, those claims are sanctioned by some other theory that you justifiably believe. Usually, the arguments for theories are going to instances of IBE: some features of actuality are taken as the data, and the claim is that they are best explained by such-and-such hypotheses. The modal claims then fall out of the relevant theories, and not vice versa.

Second, it's always possible to frame modal arguments in TEM-friendly terms. So, for example, one way to run the modal argument for dualism is to say that the modal premise follows from our folk theory of the self, and that there should be some presumption in favor of that theory. You might defend the latter claim in a number of ways. Perhaps phenomenal conservatism favors our folk theories: they seem to be true, and so we're entitled to presume that they are in the absence of defeaters. Or perhaps we should say that no competitor offers the explanatory benefits that our folk theory offers. Or perhaps there is no alternative theory that explains the precise phenomena that our folk theory explains, and so it wins by default. But whatever the defense, if it turns out that the folk theory is indeed shared, this only makes the modal argument stronger.

So much for damage control. The other option, of course, is to reject the idea that damage control is necessary. Plainly, modal arguments like the one above don't move many people. Consider an inverted version of the above argument:

1. I am a material thing.
2. If I am a material thing, then I couldn't exist without a body.
3. So, I couldn't exist without a body.

It isn't clear that the premises of this argument are more or less plausible than the premises of the initial argument. But if you think there's a difference, then your judgment about which is more plausible probably tracks whether you were previously inclined to accept the corresponding conclusion. That is, if you were already inclined to accept that I am an immaterial thing, then you probably find the premises of the first argument more plausible than those of the second, and vice versa. This explains why such arguments are often treated as enthymemes, where the suppressed premises of the first argument are:

1. I can imagine existing without a body.
2. If I can imagine existing without a body, then I could exist without a body.

The debate then shifts to these premises, which means that the original modal premise wasn't plausible enough to convince interlocutors of the desired conclusion. The upshot is that insofar as the goal of a metaphysical methodology is to make a thesis plausible to your interlocutors, this one doesn't have much going for it *when the crucial premise is an extraordinary modal claim*, as it is in the original argument.

I emphasize the last point because I am *not* saying that modal arguments have no place in metaphysics. We probably know that statues can't survive severe deformations, whereas lumps of clay can. These modal claims are sufficient to generate all sorts of puzzles—and interesting solutions to those puzzles—in the literature on the metaphysics of ordinary objects. Crucially, though, the modal claims that generate the puzzles are uninteresting ones, and TEM is perfectly compatible with the idea that some of the data for theorizing are humdrum modal facts.

To my mind, the real objection here is that TEM doesn't automatically imply that some favored argument is a good one, or that we're justified in believing some beloved metaphysical thesis. But this is as it should be. All else equal, a modal epistemology is preferable insofar as it explains why there isn't consensus about x if, in fact, there isn't consensus about x. This TEM can do. It's also plain that there is plenty of disagreement about metaphysical matters. I am perfectly happy to concede that there are people with justified metaphysical beliefs, so I'm not saying that disagreement in metaphysics undermines all justification. However, I think it's plausible that our disagreements are best explained by both the paucity of the data and the difficulty of interpreting the data available. We can put the point by saying that, while we always face the problem of underdetermination of theory by evidence, the problem is particularly acute where metaphysical theories are concerned. Hence, even if some people are justified in believing certain metaphysical theories, we probably aren't in a position to convince others that our views are true: there are too many other plausible hypotheses that are compatible with the same data. That there would be disagreement, then, is all but a foregone conclusion, just as TEM predicts.[2] So it's perfectly compatible with TEM that a favored argument is sound or a thesis true. But it is no fault of TEM if it takes additional work to establish as much.

3.5 Idealization and Partial Justification

Someone might point out that many theories idealize their target systems, and some people think that idealized models do not represent their target systems: e.g., Klein (2008). (Or if we grant that idealized models do represent their target systems in some way or other, aspects of them plainly don't—otherwise, they wouldn't be idealizations.) Relatedly, someone might worry that we're only justified in believing parts of our best theories, so that even though one of our theories says that p is possible, we aren't therefore justified in believing that p is possible.

These points complicate the story about how you move from what a theory says to what you're justified in believing, but they are just the right problems to have. It *isn't* plausible that I—with my embarrassingly poor understanding of physics—am in any position to assess what is and isn't possible for neutrinos. It takes more than

[2]It's worth considering whether we'd predict the same amount of disagreement if conceivability were our basic source of justification concerning possibility. I'm not so sure we would.

a passing familiarity with the relevant theories to make such assessments, which is why it's important for philosophers (and everyone else) to defer to those who can interpret those theories properly.

Perhaps this problem will be easier to navigate if we're justified in believing a metaphysical theory according to which all physical regularities are contingent. But the help won't be forthcoming unless we've got an argument for that metaphysical theory. I'm in no way suggesting that we *don't* have such an argument. Rather, I'm just reiterating what I've already claimed. TEM is based on the idea that our modal epistemology just falls out of other projects—namely, the cases we build for certain physical and metaphysical theories. True, it's hard to know how to interpret some theories, and for the relatively uninformed, this is an excellent reason to be cautious about making pronouncements about those modal claims that theories do and do not sanction. We don't have an objection here; we just have additional basis for epistemic humility.

3.6 The Problem of False Theories

If TEM is true, then we don't have justified beliefs about the following claims (i.e., we do not justifiably believe that they are true or that they are false):

- there are Newtonian worlds,
- wood could have contained phlogiston,
- atoms could have been indivisible,
- it could have been the case that all is water, and
- it could have been the case that the function of the brain is to cool the blood.

The reason for this is simple. We don't justifiably believe any theory that implies that these claims are possible—our best physics and chemistry and biology say otherwise. However, many people are inclined to think that these claims *are* possible, and this partly because the theories behind them had some explanatory power when they were in vogue. (Or if the above examples don't move you, substitute ones derived from other—perhaps more recent—theories that do.) What should an advocate of TEM say about this?

To begin, we should distinguish between conditional and unconditional modal knowledge. This is *not* the distinction between simple possibility and necessity claims, on the one hand, and counterfactuals on the other. Instead, you have conditional modal knowledge just in case you know that if q is possible, then p is possible. By contrast, you have *un*conditional modal knowledge that just in case you know that p is possible, full stop. TEM—like virtually all modal epistemologies—is a theory of unconditional modal knowledge, and it recommends caution about the scope of that knowledge. However, conditional modal knowledge is much easier to come by, and I suspect that it's sufficient for many purposes.

After all, much of our modal reasoning is hypothetical: it doesn't matter whether it's *really* possible that p; it just matters whether it's possible that p given certain assumptions, or on such-and-such a view, or what have you. Moreover, it's often easy to tell whether p is possible given that q is possible. If, for example, it's possible for there to be *talking* naturally pink donkeys, then it's possible for there to be naturally pink donkeys—we can just deduce the latter from the former. Likewise, if it's possible for *two* material objects to wholly occupy the same location at the same time, then it's probably possible for *three* material objects to wholly occupy the same location at the same time—if the properties of material objects don't prevent the former, then they probably permit the latter. Of course, it remains controversial whether there can be naturally pink donkeys that talk, and whether two material objects can wholly occupy the same location at the same time. But that's irrelevant here: the point is just that *given* these modal claims, we can reason from them, which is all that conditional modal knowledge requires. So, while TEM says that we aren't justified in believing that wood could have contained phlogiston, we can still reason about and from these sorts of claims, which is enough to handle cases where we are interested in the implications of discredited or not-yet-credited theories.

Are there any circumstances where we can't get by with the conditional analogues of the claims with which we began? I think that there are two common cases.

The first one involves an interlocutor who proposes a thesis to which we would like to object, but our objection involves a contentious modal claim (e.g., someone proposes the mind-body identity theory, and we object with the possibility of disembodied existence). Plainly, we're in a better epistemic position if we justifiably believe the modal claim on which our objection relies. However, we aren't in a substantially-better *dialectical* position. On the one hand, if the critique is supposed to be internal—i.e., if our interlocutor is expected to believe the modal claim—then we needn't believe the modal claim at all. We can simply point that her thesis implies (say) that p is impossible, though she believes p to be possible; that's enough to cause her some trouble. On the other hand, if the critique is supposed to be external—i.e., if we do not expect that our interlocutor will believe the modal claim—then our objection doesn't carry much weight to begin with, and we're just preaching to the choir. Insofar as our justification *is* relevant to our interlocutor, it's only insofar as it creates a worry about peer disagreement. However, our interlocutor's reasons to believe her thesis are reasons to think we're wrong about ours. So, she has reason to think we *aren't* epistemic peers. What's more, disagreement cuts both ways: if it undermines our interlocutor's justification, then it undermines ours as well. Either way, it doesn't matter much whether we have conditional or unconditional modal knowledge: the dialectic doesn't change.

The second case involves trying to develop a hypothesis about the general shape of modal reality. Given the scope of the hypothesis, we want as many data points as we can get, and a belief only qualifies as a data point if it's justified. (It's of no use to frame a hypothesis based on our guesses about what is and isn't possible.) But while this is a worthy aim, it's far from clear that we should expect to be able to develop a well-confirmed theory of the shape of modal reality. If we are realists about modality, we should expect this to be a difficult task. Moreover, there is no obvious reason why a proponent of TEM must agree with her critic about the space of worlds. If the proponent and critic disagree about what's possible, then they may also disagree about the general principles that organize the modal facts—e.g., the precise relationship between physical and metaphysical possibility, or whether any world is accessible from any other. But in the absence of an argument to the effect that our beliefs about those principles are more secure than the arguments for a particular modal epistemology—and an argument to the effect that those principles are the ones that the critic defends—it's hard to see where the difficulty lies.

So TEM requires denying that we know some possibility claims that we may be tempted to affirm. But TEM's advocate can hold his ground, especially once we recognize that our dialectical aims can often be achieved with conditional modal knowledge.

3.7 The Modal Collapse Objection

Either we have theories about what's actual, or we don't. Suppose we do. Recall that, according to TEM, a theory says that p is necessary just in case every model of the theory is a p-model—i.e., a model that satisfies p. Now let's imagine that I've forgotten where I left my keys. I consider the available information, and I form a theory about where they are: my theory is that they are on the kitchen counter. However, *every* model of this theory says that my keys are on the counter, and hence this theory says that it's *necessary* that my keys are on the counter. Alternately, suppose we *don't* have theories about what's actual. But what theories are we justified in believing if *not* theories about what's actual? The worry now is that we never satisfy TEM's first condition—i.e., we never justifiably believe a theory according to which a non-actual possibility claim, p is true—and hence have no justified beliefs about such matters whatever.

Let's work with the first horn of the dilemma. Suppose that we have a theory about the way things are in some respect: about what my grandfather did for a living, or how many honey-producing beehives there are in the U.S., or the rate at which signals are transmitted over coaxial cable. Now, by the lights of any theory, its positive claims are necessary (i.e., the claims in the theoretical definition), since they are true in each of its models (else they wouldn't be among the theory's models). So let's suppose that a theory claims that there are two and a half million honey-producing beehives in the U.S. According to TEM, if I am justified in believing this theory, then I am justified in believing the modal claims it sanctions. So, I am justified in believing that it is *necessary* that there are roughly two and a half million honey-producing beehives in the U.S. And surely it is implausible that this is so.

There are two ways to handle this problem. The first is to find a way to restrict our conception of theory so that it cannot play the requisite role in the argument. You might argue that any theory ought to be suitably general, or support interesting counterfactuals, or have a certain measure of explanatory power, or what have you. I think that this response is viable, but working out the details would be a significant and unnecessary digression, as there is an easier reply available: namely, that the

Modal Collapse Objection reduces to the (yet-to-be-discussed) Conflict Objection. I'd wager that there is a model of the logic I believe in which I run a three minute mile; however, I doubt that there is any such model of the biology I believe. In this respect—and no doubt in others—logic and biology are not in agreement about what I can do. Analogously, there is probably a model of the melittology[3] I believe in which there are only one million honey-producing beehives in the U.S., though there is no such model of my theory about what's actual. I need to say something about how disagreements between theories are to be adjudicated, and unless the advice is *always defer to the more conservative theory*, then it should go a long way toward undermining the Modal Collapse Objection. So, on to the Conflict Objection.

3.8 The Conflict Objection

Here is the Conflict Objection. There is probably no model of the biology I believe in which a human being lives to her five hundredth birthday, but I'd bet that from the perspective of the physics I believe, this state of affairs is possible, albeit wildly unlikely. What does TEM recommend in this sort of case? Should I defer to one theory or the other? If so, which one? And why? Alternately, am I justified in believing both claims, perhaps with each indexed to the appropriate 'sort' of possibility? Or, do the two theories 'cancel out', so that I am not justified in believing anything about the matter?

We can make the problem sharper still by recalling POSS and NEC, TEM's clauses about possibility and necessity:

> [POSS] For any interesting modal claim, *p*, you justifiably believe that *p* is possible if and only if (a) you justifiably believe a theory that says that *p* is possible, (b) you believe that *p* is possible on the basis of this theory, and (c) you have no defeaters for the belief that *p* is possible.

> [NEC] For any interesting modal claim, *p*, you justifiably believe that *p* is necessary if and only if (a) you justifiably believe a theory that says that *p* is necessary, (b) you believe that *p* is necessary on the basis of this theory, and (c) you have no defeaters for the belief that *p* is necessary.

TEM gets into trouble here because POSS and NEC either (1) allow cases where you are justified in believing both that it is possible that *p* and that it is necessary that $\sim p$ or (2) lead to the conclusion that you are not justified in having either belief, since each constitutes a rebutting defeater for the other. This is the Conflict Objection.

---

[3]Melittology is the study of bees; apiology is the study of honeybees specifically. Of course, my knowledge of bees of any kind is very limited; if I know anything beyond the platitudes about bees (they are black and yellow; they can sting you; they make a 'bzzzzing' sound; etc.), then that's not a result of knowing much in the way of theory, but only because of information gleaned from news articles about them. So most of *my* modal knowledge about bees is parasitic on the knowledge enjoyed by experts, via reporters. I expect that this is the normal state of affairs, and I doubt that it is good news for the scope of our modal knowledge. For more on this problem, see below.

A natural first step is to introduce kinds of possibility and necessity. The new version of NEC might go as follows:

> [φ-NEC] For any interesting modal claim, *p*, you justifiably believe that *p* is φ-necessary if and only if (a) you justifiably believe a theory that says that *p* is φ-necessary, (b) you believe that *p* is φ-necessary on the basis of this theory, and (c) you have no defeaters for the belief that *p* is φ-necessary.[4]

We now face the question: under what circumstances will a theory say that *p* is φ-necessary rather than necessary *simpliciter*? We can think about this in terms of the theory's generality. If a theory purports to apply to *all* phenomena, then if it says that *p* is necessary, it says that *p* is necessary *simpliciter*. However, if a theory limits itself to *physical* phenomena—or biological phenomena, or terrestrial biological phenomena, or carbon-based terrestrial biological phenomena—then it's modal claims are restricted accordingly: the theory won't say that *p* is necessary *simpliciter*, but rather that *p* is physically necessary, or biologically necessary, or what have you. These 'types' of necessity are just restrictions on the space of possibilities: after you set aside the non-φ-worlds—i.e., the worlds where the actual φ-laws[5] don't hold—φ-necessity is just necessary *simpliciter* evaluated just with respect to the φ-worlds.

Now the Conflict Objection goes as follows. *p* is φ-possible according to Theory #1; ~*p* is ψ-necessary according to Theory #2. We want to know whether *p* is possible or impossible *simpliciter*. What should we believe? Let's make this concrete. Some botanists think that it is biologically impossible to have genetically engineered (GE) crops adjacent to non-GE crops without cross-contamination within a single generation.[6] I take it, therefore, that there is no model of the biology

[4]N.B., there are actually two ways to interpret this version of NEC. On one, the theoretical definition is less specific, but the theoretical hypothesis indicates that only certain models are to be taken as representing possibilities—namely, the ones that satisfy certain assumptions about the logic in play, or the physical environment, or what have you. On the other, this information is built into the theoretical definition from the outset. For my purposes, the choice between these options is largely arbitrary.

[5]I don't think that a commitment to φ-necessity requires a commitment to laws, though the two obviously sit well together. You can get the same result by saying that φ-necessity is truth at all the worlds of which the final φ-theory is true (i.e., biological necessity is truth at all of the worlds of which final biology is true). You might also be able to get the same result by saying that φ-necessity is truth at all the worlds that are relevantly similar to the actual world, where the relevance relation is cashed out in terms of particular functional relationships, or systems, or mechanisms, or whatever (depending on the value of φ).

[6]Or, at least, that's the way they talk. If pressed, they might say that it's *extraordinarily unlikely*—and so impossible for all practical purposes—that cross-contamination not occur. But then I could make the point in terms of practical possibility. In any case, if it's *true* that it's extraordinarily unlikely, then this has significant ramifications. Monsanto, for example, is a huge biotech firm that owns the rights to a number of GE seeds. Many, many U.S. farmers use these seeds, and a few years ago, Monsanto made the news for suing the smaller farmers who haven't gotten on board. The problem is that these smaller farmers own fields that are adjacent to ones that use Monsanto seed. As a result of cross-pollination, the smaller farmers end up with GE crops for which Monsanto has the patent. As a result, the small farmers are forced to destroy their seeds at the end of each

they believe in which GE crops are adjacent to non-GE crops without the relevant contamination. However, I also take it that this is because their models assume certain statistical facts about wind patterns, the behaviors of insect and animal populations, and so on. From the perspective of physics, though, I take it that there is no reason why cross-contamination must occur. It is wildly unlikely, of course, that pollen from one group of plants wouldn't make its way to an adjacent group, but in an indeterministic framework, nothing *rules out* this possibility; the thing to say is just that the odds are very, very low that it will be actualized. But that, of course, is to agree that the possibility is genuine. On the assumption that I've correctly described the situation, the question now is this: is it or is it not possible *simpliciter* that GE crops co-exist alongside non-GE crops without contamination? In this case, I see no reason to defer to the biological theory; the botanists make plausible assumptions for their purposes, but they are assumptions that needn't be true, as the physical theory shows. So, it seems to me that co-existence is possible *simpliciter*.

At this juncture, you might wonder whether we should *always* defer to the more general theory. The answer, in short, is that you should—unless there are specific reasons to doubt its deliverances. (Logic might be just this sort of exception, as I'll soon discuss.) But this point shouldn't be taken to license too much optimism when it comes to the scope of our modal knowledge—or at least *my* modal knowledge. In the above case, I suggested that our physical theory has a model in which grains of pollen do not make it from one field to another for an entire growing season. In truth, though, I have no idea whether this is the case. First of all, my grip on the relevant physics is pretty weak. For virtually all intents and purposes, I have a Newtonian conception of the physical world, chastened only by a concern to agree with the things said by people who actually understand quantum mechanics. Moreover, because the relevant physics is obscure to me, the *application* of the relevant physics is obscure to me. When I reason about these matters, I assume that the upshot of indeterminism is that just about any momentary or sustained configuration of objects is possible, though most are wildly improbable. But I certainly don't have an argument for this, and for all I know, the thesis of indeterminism is supposed to apply only to certain microphysical particles, and not to things like grains of pollen. Second, insofar as I even *believe* the relevant physics, I probably believe it second-hand, which is to say that I believe it in virtue of having the disposition to defer to experts in my community on physics-related matters. If so, though, then my modal judgments are not really based on *physics*, but rather on the cartoon version of physics that I've pieced together from various (more or less reliable) sources. And I have a readymade defeater for any such judgment: since I have only the weakest grip on the real theory, I have only a very limited ability to tell where my cartoon version deviates from it. So, where my confidence about the parallel is low, I should be accordingly cautious in my modal judgments. In my view, this defeater is an ineliminable part of living in a society where there is such a significant division of

season—lest they violate patent law—which means that Monsanto is making small-scale, non-GM farming prohibitively expensive.

cognitive labor. In any case, the essential point here is that it isn't nearly so obvious that I believe a physical theory that has the relevant model—or that if I do, that I'm also justified in believing that the model represents a genuine possibility.

No doubt there are metaphysical theories that I might invoke to get the co-existence result, and in a different context, it would be interesting to explore what they are. Here, though, let it suffice to say this. Whatever the package of metaphysical views, the task is to show that I am justified in believing it. And to state the obvious, it's hard to provide a compelling argument for any metaphysical view, even if the members of the target audience are (the mythical) disinterested third parties to the debate. Granted, providing a compelling argument may not the only way to justify belief in a theory, but I would think that one is needed here. After all, if metaphysicians have shown nothing else, it's that wildly different theories can be made to agree with common sense. This makes it much less plausible that we can rely on common sense to determine the metaphysical theory that we believe. To be clear, I am not a skeptic about metaphysical knowledge. My point is just this: it isn't *obvious* that I am justified in believing that GE crops can co-exist alongside non-GE crops without contamination. I know that I don't understand the relevant physics, and it isn't at all clear that I'm justified in believing a metaphysical theory that specifies the relevant model. If TEM is true, then I am justified in believing that *p* is possible *simpliciter* if there is a *p*-model specified by the most general theory that I justifiably believe. In any given case, though, we need to press to see whether (a) I actually believe the relevant theory and (b) I am justified in believing it. The thrust of the last few paragraphs is that we shouldn't be blithely optimistic that *I* satisfy (a) and (b)—and insofar as I am not an outlier, we shouldn't be blithely optimistic that others do either.

3.9 The Logic Objection

The Modal Collapse Objection amounts to the worry that TEM lets in too few possibilities. That objection reduces to the Conflict Objection, which I addressed by saying you should defer to the most general theory you justifiably believe. But this solution invites a new and very serious challenge. Either the logic I believe counts as a theory, or it doesn't. If it does, then TEM is at risk of implying that I'm justified in believing that whatever is logically possible is possible *simpliciter*. In other words, it seems that I am justified in taking consistency to be a guide to possibility. But that does not seem to sit well with the considerations that led us to this modal epistemology: I've claimed that TEM is cautious, but this would make TEM into a view that gives us extensive modal knowledge—much more than most modal epistemologies.

I see the logic objection as the most serious challenge to TEM. If it goes through, then TEM is entirely trivial. So, I'm going to argue against this objection at some length. I don't just want to contend that it fails, but also that it can't be reworked. My claim is that *no* version of it is likely to succeed.

3.9.1 A False Start

To be clear, you can't reply to this objection by saying—as van Inwagen (1998) does—that there is no presumption in favor of possibility. That is, the Logic Objection is not the claim that, since certain possibilities aren't ruled out by logic, we're entitled to presume that they're possible. Rather, the charge is that *because* TEM alleges that our only guides to interesting modal matters are our best theories, and because I replied to the conflict objection by saying that we ought to defer to our most general theory, TEM is at risk of saying that our best logic is our guide to what's genuinely possible. If it's possible according to that logic, then we're justified in taking it to be absolutely possible, whatever our other theories (physical or metaphysical) might say.[7]

Russell and Frege had a universalist conception of logic: they regarded it as the most general theory, the theory that applies to absolutely everything. I would prefer not to give up this view of logic; I'll grant that our best logic qualifies both as a theory and as something that we justifiably believe. So, I need to find ways to block the suggestion that, if TEM is true, then consistency is a perfect guide to possibility. And I think there are such ways available. Logic may well be the most general theory, but this alone is not enough to show that TEM leads to the view that consistency is a perfect guide to possibility. To get there, you also need the assumption that, for any given possibility claim sanctioned by our logic, we have no defeaters for it. Is that the case?

I don't think so, but it's important to recognize how *not* to defend the existence of defeaters. A tempting move here is to appeal to claims that (a) seem to be necessary and (b) are such that their denials aren't contradictory. One classic example involves color: nothing can be red all over and green all over simultaneously. That seems to be necessarily true, yet it isn't contradictory to suppose that something *is* red all over and green all over simultaneously. Plainly, though, that move won't work with TEM. To justify the modal claim, you'd need a theory of color according to which nothing can be red all over and green all over simultaneously. But once you have the theory, the problem comes roaring back, since we now have to explain why the theory about color isn't superseded by our best logic.[8]

---

[7]Granted, there are other interesting issues in the neighborhood. If logic doesn't count as a theory, then what is it? How are our logical beliefs justified? This is especially pressing, since for any theory, we need know whether it is consistent in order to know whether it has any models at all, and hence we need some justified logical beliefs for TEM to get off the ground. Finally, *which* logic are we justified in believing? If more than one, then for which purposes should we deploy which logic? I bracket all these questions here.

[8]Similar points could be about other property incompatibilities: e.g., being God and not knowing the past, or being a panther and being reptilian, or being *this* clay pot (right here) and being made of *that* material (over there). If these propositions are indeed necessary, their necessity does not seem to be explicable in terms of their logical form. However we would like to classify this breed of necessity, it looks as though consistency outstrips it, since there are consistent but metaphysically impossible propositions (e.g., *something is simultaneously red all over and green all over*).

Given the failure of this move, I want to develop three others in increasing detail. The first is the most ambitious: it involves arguing directly for the view that the logical possibilities outstrip the metaphysical or genuine possibilities. The second two are more modest: they involve granting that if *p* is a logical possibility, then it's genuinely possible, but then denying that this is of much epistemic significance.

3.9.2 *The Ambitious Option*

The ambitious option is to argue that we have theoretical reasons to deny that logical possibility is coextensive with genuine possibility.

Alexander Pruss (2015) gives an argument to this effect. As he points out, you can consistently deny that that Robinson's arithmetic is consistent—that much follows from Gödel's Second Incompleteness Theorem. (See Hájek and Pudlák 1998). But it also seems that Robinson's arithmetic is either necessarily consistent or necessarily inconsistent. The consistency or inconsistency of a set of axioms isn't the sort of thing that we'd expect to vary across worlds. Given the plausible assumption that Robinson's arithmetic *is* consistent, it follows that it's impossible for it to be inconsistent, which means that consistency doesn't imply possibility.[9]

However, it's tempting to dismiss Pruss's argument by saying that mathematics somehow constitutes a special case. Granted, there *is* a gap between consistency and possibility, but that gap doesn't matter when it comes to metaphysics: consistency is still a good guide to possibility there.

This doesn't seem to be right either. Some metaphysicians, after all, have independent reasons to reject logical possibility is coextensive with genuine possibility. Platonists, for example, might want to say that we have good reason to posit abstracta and to maintain that it's genuinely necessary that they exist. However, they might also concede that it isn't logically necessary that abstracta exist. (It isn't straightforwardly *inconsistent* to deny their existence.) So, their theory about abstracta might lead them to reinterpret their logic, allowing that some logical possibilities aren't genuinely possible.[10] There are two challenges here. The first is to run the argument for abstracta; the second is to argue that their existence would be genuinely necessary. The former might not be so hard: even Quine thought there were sets, which means that the case for Platonism doesn't presuppose an inclination toward ontological excess. So, let's focus on the latter.

At first blush, anyway, you might think that it isn't worth positing abstracta unless their existence is genuinely necessary. Typically, arguments for abstracta involve the idea that abstracta are supposed to serve as the truthmakers for certain necessary

---

[9] As Anand Vaidya (2016) argues, this raises serious troubles for a certain (common) form of modal rationalism.

[10] We might find some theists pressing a similar line, at least if they take it that God exists necessarily but that there is no contradiction in supposing that God doesn't exist.

truths, such as mathematical truths. If they don't exist in some logically possible world, then even the mathematical truths at that world aren't true—e.g., it won't be true that $2 + 2 = 4$. So, if we allow that there are genuinely possible worlds at which abstracta don't exist—a state of affairs that seems to be logically possible—then we'll have to allow that there are worlds at which certain necessary truths aren't true. Since we shouldn't allow that, we shouldn't say that every logically possible world is genuinely possible.

Someone might object that this line of argument is based on deriving a logical impossibility from the non-existence of abstracta: namely, certain necessary truths not being true. So although it might *seem* as though it's logically possible for abstracta not to exist, it isn't *really* logically possible. In other words, if the argument succeeds in showing that we shouldn't postulate worlds without abstracta, it's because the argument shows that their existence is logically necessary.

Not so. The argument I sketched assumes that abstracta are supposed to serve as the truthmakers for certain necessary truths. There is no obvious contradiction in saying that those truths don't have truthmakers, or maintaining that they have concrete truthmakers, or what have you. The assumption about the role of abstracta is a hypothesis that's supposed to explain various features of mathematical truth, but it could well be wrong (in the sense that it isn't inconsistent to say it's wrong). The question, of course, is whether the initial argument for abstracta succeeds—an argument that's predicated on this assumption about what abstracta do. If so, then we can avoid the conclusion that consistency is a perfect guide to possibility.

It's pretty clear, though, that it would take us far afield to pursue this matter in any real detail, and at any rate, it's unlikely to convince any nominalists. How else might we resist the logic objection? Well, even if consistency is coextensive with absolute possibility, consistency is not of much epistemic use. That is, while consistency might be a perfect guide to possibility, it might not be a perfect guide for *us*, given our epistemic limitations. There are two routes to this conclusion.

3.9.3 A Posteriori Necessities

The first route relies on two suppositions: on the one hand, that we can establish identity claims based on a posteriori considerations; on the other, that we somehow know that identities are necessary. (The latter claim is easy to establish given TEM: we are justified in believing standard first-order logic with identity, and that theory has no model in which $a \neq a$; therefore, we are justified in believing that identities are necessary.) We now have the old epistemological worries created by a posteriori necessities. To be clear, a posteriori necessities do not create trouble because they lack the requisite form[11]; rather, they are awkward because the form

[11]...or fail to entail a proposition with the requisite form. I'll suppress this alternate formulation in what follows.

of the proposition is not obvious from the form of the sentence that expresses it. The sentence < Water is not H_2O > may indeed express a proposition that has the form "$a \neq a$,"[12] but if you don't already know this, then the sentence won't tip you off. (As Nathan Salmon (2007) would put it, you believe the proposition under a guise that does not presume the identity of water and H_2O; hence you have no cognitive access to (the relevant aspect of) the form of the proposition.) The upshot is that if you are judging consistency based on the form of the sentence, then your guide to possibility is highly fallible: it's going to deliver false positives whenever you are unaware of the relevant empirical facts. And given (a) the sheer complexity of the natural world and (b) the division of cognitive labor that characterizes modern society, there should be little doubt that you are unaware of many such facts. (Parallel problems crop up with property identities if we allow second-order quantification.) The obvious way to guard against this difficulty is to limit your use of this guide to those cases where you have positive empirical reasons to believe that $a \neq b$, not merely the absence of evidence that they *are* identical. What would constitute having such reasons? First and foremost, having theories about the natures of a and b. But this, of course, is a much-chastened version of the view that consistency is a guide to possibility, and hardly constitutes a threat to the caution that TEM demands.

3.9.4 Semantic Indeterminacy

What's the second route to the conclusion that consistency isn't of much epistemic use? In short, the thought is that even if p seems to be consistent, that consistency may well be a product of semantic indeterminacy. In such cases, we shouldn't rely on consistency as a guide to possibility.

It will take some conceptual machinery to make this point clearly, as I want to avoid a number of potential objections. To begin, let's think about the meanings of singular and general terms as metalinguistic rules. These rules characterize the correct uses of those terms within the object language. (Feel free to import your favorite theory of the source of linguistic correctness, since nothing turns on it here.) Among other things, a term's rule states its application and coapplication conditions, the former being those conditions under which it can be used at all, the latter being those conditions under which it can be used in expressions like "x is identical to y." Restricting our attention to application conditions, a first pass at the rule for "house" might go like this: "house" is correctly applied just in case it's applied to a physical structure that was either constructed or modified for habitation. Of course, we don't need to have explicit knowledge of these rules in order to use the terms correctly.

[12] This is too simple, since it isn't obvious that "water" or "H_2O" are singular terms, each of which refers to the same massive scattered object. However, I don't think that the details matter for my purposes here.

The point is just that correctness is defined in terms of consistency with the rules, which are themselves determined by various factors—e.g., certain psychological, social, and historical facts.

These rules determine the "frame-level" application and coapplication conditions. In other words, and again focusing on application conditions, they specify that, if the term refers at all, it refers to an object (or class of objects) satisfying some more general description. (In my toy example, the frame-level application condition for "house" is the bit requiring that it refer to a physical structure.) If terms have frame-level application and coapplication conditions, they can be used to ground analytic entailments. Analytic entailment is a metalinguistic relation that is, roughly, something like sufficiency for truth. Suppose that a sortal, φ, has been correctly applied to some object x, and the sortal's associated rule states that it applies only if x also falls under some other sortal, ψ; in that case, "x is φ" analytically entails "x is ψ"—i.e., "x is φ" is sufficient for the truth of "x is ψ." Similarly, suppose that a sortal, φ, applies just in case conditions C_1—C_n are satisfied; then, "C_1—C_n are satisfied" analytically entails "there is a φ" (or "x is φ," depending on the sortal and the circumstances in question), and vice versa. Analytic entailments are also necessary entailments: if you know that "x is φ" analytically entails "x is ψ," then you know that, necessarily, if x is φ then x is ψ. And if you know that "x is φ" does *not* analytically entail "x is ψ," then you know that, possibly, x is φ but not ψ. Because analytic entailments are grounded in the rules that govern the uses of terms, competent speakers and reasoners should be able to know that analytic entailments hold simply by knowing the meanings of the terms involved. The upshot? Our justification for our modal beliefs can be grounded in linguistic and logical competence, just as the Logic Objection supposes.

As I said, analytic entailment is supposed to be a metalinguistic relationship, which is to say that it's supposed to be a relationship among propositions, and therefore only derivatively a relationship between sentences or utterances or inscriptions.[13] Note, moreover, that if there are no determinate relationships among propositions, then there are no analytic entailments: for any two propositions Φ and Ψ, if it is indeterminate whether Φ analytically entails Ψ, then it isn't *true* that Φ analytically entails Ψ.[14] However, we can ignore this issue here. What

[13]Not everyone would agree with this characterization of analytic entailment, since analytic entailment is a species of entailment, and not everyone thinks that entailment holds among propositions. However, most of the parties to this debate are happy to talk in terms of entailment among propositions, so I will not defend the assumption.

[14]Note that your theory of propositions will determine how you spell out relationships among propositions. If you think that propositions are, say, functions from worlds to truth values—as Robert Stalnaker does—then you'll have a very simple explanation of when one proposition, Φ, analytically entails another, Ψ: Φ analytically entails Ψ iff the set of worlds to which Φ assigns the value "true" is a subset of the set of worlds to which Ψ assigns the value "true." But if you think that propositions are structured entities—as neo-Russellians think—then you'll have to give some other account. The crucial point for present purposes is that different conceptions of propositions have different implications concerning the nature of relationships among propositions. On Stalnaker's view, it's easy to see how relationships among propositions might be determinate or indeterminate:

matters is the fact that, regardless of your view of propositions, you have to explain how sentences (at times in contexts) are correlated with the propositions that they express. It's only if those correlations are established that you can have derivative analytic entailments. What I'll argue is that indeterminacy probably infects the meanings of the terms used to express claims about extraordinary modal matters. If so, then when a sentence contains one of those terms, it's indeterminate which proposition the sentence expresses. So, even if the relationships among propositions *are* determinate, there won't be any fact of the matter as to whether one sentence containing an indeterminate term derivatively analytically entails another.[15] And if that's right, then we'll have a defeater for inferences from consistency to possibility. Those inferences aren't simply based on what our logic says, but on the meanings of the non-logical terms. If those meanings are indeterminate, and any of the plausible resolutions of the meaning preclude the possibility in question, then we shouldn't make the inference until we know how that indeterminacy ought to be resolved.[16]

To make this point more clearly, let's consider a simple case. Suppose we know, of a particular individual, that he's male and that he's unmarried. That analytically entails—and we can therefore know—that he's a bachelor. We'll consider the issue of semantic indeterminacy in a moment, but first let's think about why you might question the idea that the truth of "he's male" and "he's unmarried" is sufficient for the truth of "he's a bachelor." Marcus Giaquinto offers two cases that seem to have precisely this implication:

1. Peter once married a refugee merely to save her from deportation to a country suffering civil war; after the marriage ceremony he never saw her again, but continued to live as young single men do, available for a long-term personal partnership.
2. Mike [...] never married; he and his partner have been together for several decades, have two adult daughters and continue to live together as companions (Giaquinto 2008, 97).

---

the issue turns on whether every proposition is defined for every world. On neo-Russellian views, it's a bit harder to see how the details will go. If the constituents of propositions cannot be vague, then, presumably, all relationships among propositions will be determinate. However, if vague objects can be constituents, or if even senses themselves can be constituents, then perhaps some story could be told on which those relationships would not be determinate. So, if we were being careful, then we'd say that skepticism about analytic entailments commits you to skepticism about determinate relationships among propositions, which commits you to certain views about the nature of propositions.

[15] ...at least where the semantic value of the indeterminate term is supposed to be what explains the holding of the analytic entailment. So, for example, "That bald man owns a triangle" analytically entails "That bald man owns a polygon," but while "bald man" occurs in both sentences, that term is not what explains the holding of the analytic entailment. To put the point analogically, the occurrences of "bald man" are background conditions, not the cause.

[16] Note, moreover, since no one thinks that we have unmediated access to propositions, this problem will crop up for any view that bases our modal knowledge on the objects of our immediate semantic knowledge—i.e., the objects, whatever they are, that have meanings or express propositions, but are not themselves meanings or propositions.

I'm inclined to agree with Giaquinto that Peter, though married, is a bachelor; Mike, though unmarried, is not. But if Giaquinto is correct, then it's false that the truth of "he's male" and "he's unmarried" is sufficient for the truth of "he's a bachelor." At the very least, you need an additional proposition to the effect that the man is not cohabiting with a woman with whom he is in a long-term, committed relationship.

Just about everyone who is sympathetic to analyticity thinks that the sentence, "Bachelors are unmarried males," is a paradigmatic instance of an analytic truth. If those philosophers are wrong about a paradigmatic and relatively simple case, as Giaquinto suggests, then we should have very little confidence that we can determine sufficient conditions for the more complex cases: e.g., "baseball," "water," "pain," "red," "God," "gratuitous evil." So, even if there are analytic entailments among propositions, they may not be of much value for epistemic reasons, since we may be extraordinarily bad at ferreting them out. (The other option, of course, is that there may not *be* very many analytic entailments.)

The way to mitigate the force of this challenge is to generate a list of the reasons why conceptual analysis is hard, the upshot being that, while it's no easy task to settle the meaning of a term, there is nothing other than the requisite effort standing between us and the knowledge analytic entailments. But this is implausible, and here is where the worry about indeterminacy comes to the fore. We can represent the two candidate meanings of "bachelor" with two sets of properties: < *maleness, being unmarried* > and < *maleness, being unattached* >.[17] Now our interlocutor must answer the following question: has the meaning of "bachelor" changed, or has it remained constant? On the first view, you say that < *maleness, being unmarried* > once represented the meaning of "bachelor," but now it's best represented by < *maleness, being unattached* >, perhaps because the meaning of "bachelor" was fixed by certain social structures and mores—e.g., we used to identify bachelors based on marital status, but now we do so based on lifestyle. On the second view, you say that "bachelor" was always best represented by < *maleness, being unattached* > and speakers were incorrect if they said otherwise. It was reasonable to assume that being an unmarried male was necessary and sufficient for being a bachelor when the class of unmarried males was coextensive with the class of bachelors, but after seeing that those classes are not necessarily coextensive, we revised our belief.

Our interlocutor needs to explain why one of these responses is preferable to the other. If he can't, then we've got no reason to choose one of the two candidate semantic values. (This underscores the point that it won't do to maintain that one of these responses *is* preferable to the other, though we aren't in a position to know which.) This is no easy task. It's unlikely that any of the descriptive content that might have been employed to coin "bachelor" settled whether its meaning is best represented by < *maleness, being unmarried* > or < *maleness, being unattached* >. After all, the odds are good that the distinction wasn't at the forefront of anyone's mind when the term was coined. Alternately, perhaps the issue wasn't settled

[17]To be charitable, let's assume that these are the only two candidates and that we're right about what the candidates are. Also, note that I'm suppressing *adultness* for ease of exposition.

then, but it's settled now. But when did we precisify the meaning? And how? It certainly seems as if the ambiguity is still present, as evinced by the appeal of both candidate meanings. It's far more plausible that the meaning neither changed nor was precisified: the meaning was and still is indeterminate. Hence, there never was, and still is not, a fact of the matter concerning which of the two sets represents the "real" meaning of the term, on the one hand, and which is the imposter, on the other.

Why suppose that this a general problem, and not one that merely afflicts "bachelor"? Well, one reason is that, if a term like "bachelor" suffers from semantic indeterminacy—which was supposed to be the paradigm case of an analytic truth—then we would expect it affect non-paradigm cases. That there is ambiguity in an (ostensibly) easy case casts doubt on the harder ones.[18]

Another reason, and the more important one, is that terms are inventions. Consequently, we have little reason to suppose that their meanings are significantly more precise than are required by the ends for which we use them. After all, the point of having terms is to divvy up the things in the world. It would be startling if this need led us to generate meanings for those terms that are precise enough to allow us to decide what they require in all circumstances, since our circumstances rarely require such precision. (Linguistic evolution, like evolution generally, is satisficing, not optimizing.) Most people's circumstances, for example, have never required the semantic precision needed to answer questions like these: How much hair can you have on your head while still being bald? Is an El Camino a car? If you breed foxes until they become very docile and look like dogs (a feat that you can accomplish in about forty generations), are they dogs? If a transgender man (i.e., someone who was identified female at birth, but now identifies as male) is unmarried, is he a bachelor? Is water H_2O, or just one *species* of H_2O—namely, namely, the one *without* extra deuterium (which makes for heavy water; or without protium, for semiheavy water; or without tritium, for tritiated water)?[19]

Granted, we could refine the relevant concepts so that there would be answers to the above questions. But precisification is *not* uncovering existing detail; rather, it is *adding* detail where none existed previously. Moreover, it isn't clear how and when precisification happens. Just because some scientists begin to talk a certain way about water, and we follow suit, that doesn't show that the meaning has changed. It's equally likely that we're just deferring to scientists concerning *truths* about water. On such a view, "water" needn't analytically entail "H_2O"; rather, water's being H_2O might just happen to be among the truths that we all take for granted about water.

---

[18]On that note, it's plausible to think of vague predicates ("bald," etc.) as having indeterminate semantic values, and natural languages are shot through with vague predicates. As with "bachelor," the candidate semantic values for vague predicates may not be disjoint. That is, if some property φ is shared by all the candidate semantic values for some predicate ψ, then "x is ψ" analytically entails "x is φ." Some contest this, arguing that the candidate semantic values usually *are* disjoint (or, more carefully, that there usually are pairs of disjoint candidate semantic values for a given term); see the arguments in LaPorte (2004).

[19]For more on this point, see Weisberg (2006).

The upshot is as follows. When *p* seems to be consistent—i.e., when there are no analytic entailments that seem to preclude *p*'s possibility—there are (at least) two explanations for this. The first is that *p* includes terms with determinate meanings and is in fact consistent. In that case, it's logically possible. But especially when considering extraordinary modal claims (though not only then), a second explanation will tend to be more likely. According to this second explanation, *p* seems to be consistent because the terms used to express *p* don't have determinate meanings. On some precisifications of the meanings of the terms used to express *p*, there are no analytic entailments that preclude *p*'s possibility. On others, however, there are. And until we know how to precisify those terms, we should suspend judgment about *p*'s consistency.

Some might think that this is a pseudo-problem, since we can simply distinguish the semantic candidates and assess whether there are analytic entailments that hold under the various precisifications of the terms. But while we certainly can distinguish the semantic candidates, that's of no use here. If I want to know whether there could be naturally purple cows, it doesn't matter whether there is a way of construing that statement such that it expresses a consistent proposition. Rather, I have a question about whether cows can be naturally purple, and I don't want to lose *that specific question*—which is about whether particular beings can have a particular property in a particular way—by refining the terms so that it expresses a more precise proposition. After all, *that* proposition won't be about cows, purpleness, and naturalness *as I currently use those terms*, but about cows*, purpleness*, and naturalness*. And until I have some reason to precisify my terms accordingly, I should be agnostic about whether that proposition is the one I wanted to examine.

The question now is: how should we precisify our terms? I think the most plausible answer involves revisiting the considerations that led us to use that term in the first place. Given what we now know, how should that term evolve? Answering that question will probably require balancing various explanatory and pragmatic factors. What does the term do for us? What phenomena is it designed to capture? Does the term reflect our current understanding of that phenomena, or is it deficient in some way or other? Can any of those deficiencies be justified on pragmatic grounds—e.g., would the precision make the term less useful given how it's typically used? If so, then how should that fact affect the inferences we make in cases well beyond the normal application of the term? We can, of course, work through these questions in particular cases, refine our terms, and use the presence or absence of analytic entailments to make judgments about what's logically possible. But at that point, we'll be operating in a way that isn't far from what TEM recommends. That is, we'll be precisifying the meanings of terms based on what we know—or what we take ourselves to know—at the time. And at least in part, that will involve relying on our best theories. So, though there is a way out of semantic indeterminacy, it's a way that mitigates the force of the Logic Objection.

3.9.5 *Summing Up*

We've now explored three replies to the Logic Objection. The first involved rejecting the idea that consistency is coextensive with possibility, either via Pruss's (2015) argument, or via considerations relating to abstracta.

The second reply involves granting that the logical possibilities are coextensive with the genuine possibilities. However, this reply maintains that consistency is not of much epistemic use because of a posteriori necessities. Given TEM, it's plausible that we're justified in believing that identities are necessary. Given the standard battery of Kripkean arguments, it's plausible that the meanings of some natural kind terms are their referents, and they rigidly designate the same items in all worlds in which they refer. Given a natural interpretation of what scientists tell us, it's plausible that they have established some identities. So, we're in the position of not always knowing which proposition we're expressing when we utter a sentence—we can't know whether we're running afoul of some a posteriori necessities.

The third reply also involves granting that the logical possibilities are coextensive with the genuine possibilities, and it also insists that consistency is not of much epistemic use. However, it gets to that conclusion by appealing to semantic indeterminacy, which undermines the claim that apparent consistency is based on the right factors.

It seems to me, therefore, that the logic objection fails. Consistency isn't coextensive with possibility, and even if it were, consistency wouldn't be a good guide to possibility.

It's now clear what TEM says, and with all these objections behind us, it should also be clear that TEM is a viable approach to the epistemology of modality. Demonstrating viability, however, is a long way from demonstrating that the view is worth adopting. Making the latter case is the task of Part II.

Part II

Chapter 4
Theory Selection

Abstract This chapter begins Part II. In it, I develop a general framework for selecting between competing modal epistemologies, focusing on three desiderata that are especially relevant to our choice: conservatism, simplicity, and the ability to handle modal disagreement. Unsurprisingly, I argue that TEM fares well in the relevant respects. However, I'll also distinguish the senses in which TEM is and isn't conservative, eventually describing TEM's disagreement with some modal intuitions as its cautiousness. This requires further discussion in subsequent chapters.

As I said, this book has two parts. The last three chapters composed the first: in them, I argued that TEM is a viable standard to which we could hold interesting modal claims. The remaining chapters compose the second. In essence, my aim in Part II is to show that TEM has a number of features that make it worthy of acceptance.

In this chapter, I develop a general framework for selecting between competing modal epistemologies, focusing on three desiderata that are especially relevant to our choice. Unsurprisingly, I argue that TEM fares well in the relevant respects. However, I'll also raise the issue of TEM's cautiousness, which requires further discussion in subsequent chapters.

4.1 How Should We Choose?

In the 1970s, modality made its comeback. Its epistemology wasn't far behind. And some 40 years later, accounts of modal knowledge are many and varied, providing a bewildering array of options to any newcomer.[1] How should we choose between them? If we go by the literature, it may seem as though there is little consensus on this matter. Some seem concerned to preserve our intuitions about

[1]See McLeod (2005), Evnine (2008), and Vaidya (2015) for overviews of the literature.

B. Fischer, *Modal Justification via Theories*, Synthese Library 380,
DOI 10.1007/978-3-319-49127-1_4

cases; others don't (e.g., Yablo (1993) vs. van Inwagen (1998)). Some argue for a theory based on an antecedent metaphysical framework; others don't (e.g., Peacocke (1999) vs. Kung (2010)). Some appear to be motivated by more general epistemological principles; others look for principles unique to the domain (e.g., Biggs (2010) vs. Rasmussen (2013)). But this diversity is only apparent. The modal epistemology literature isn't characterized by brute disagreement about method: rather, it's characterized by attention to different desiderata. (If it were otherwise, we'd expect to see people arguing that alternative theories are unmotivated. But that charge rarely appears.) What we need, therefore, is some clarity about what the desiderata are and how we ought to employ them.

My aim in this chapter is to articulate some of the general methodological considerations that should guide our selection. I do this based on two assumptions: first, that we should understand modal epistemologies as competing explanatory hypotheses; second, that we should use inference to the best explanation (IBE) to decide between them. Before going further, I should make two points. First, while I'm registering these claims as assumptions, I don't mean to suggest that there is nothing to be said in their defense. We should understand modal epistemologies as competing explanatory hypotheses because, on the face of it, we offer them to explain philosophical data: the distribution of justified and unjustified beliefs, the varying degrees of justification that our beliefs enjoy, why certain disagreements seem to be intractable, why certain apparent defeaters are only apparent, and so on. And we should employ IBE to choose between them because we have no real alternative. Perception is of no help here, there is no sample from which to make an inductive argument, and it isn't an a priori truth that some particular modal epistemology is the best one. Moreover, it would be irresponsible to endorse a modal epistemology *just* because it fits well with one's preferred semantic theory, or *just* because it implies that some theory of mind is correct, or *just* because it is intuitive or elegant or otherwise appealing. To take any such path would be to ignore the many other factors that are plainly relevant—not least, fit with all our *other* commitments. At least in this case, holistic evaluation is the only game in town. Given that the evaluation is of explanatory hypotheses, IBE is the obvious framework with which to begin.

The second point is this: as will be plain from what follows, we aren't settling much by deciding to use IBE. Its usefulness consists in the structure it provides for *raising* questions, not dodging them. Our two assumptions allow us to think of the desiderata in modal epistemology as being akin to the explanatory virtues—i.e., the criteria for assessing the best explanation. Then, we can use debates about the merits of IBE to be more precise about the significance of various desiderata, as well as how they might be used in tandem to guide our selection.

In the next section, I discuss IBE in more detail, explaining how we can use it to illuminate issues in modal epistemology. The following two sections are the heart of the chapter. In the former, I map out the desiderata; in the latter, I suggest how they should be employed jointly.

4.2 IBE

Let's begin with a fairly standard formulation of IBE:

1. Hypothesis h would, if true, adequately explain facts f_1—f_n.[2]
2. h would, if true, explain f_1—f_n better than any of its competitors.
3. So, probably, h is true.

This formulation raises two issues. First, in what sense is h better than its competitors? Second, what's the appropriate attitude to have toward the argument's conclusion?

The classic answer to the first question is that h fares better than its competitors on the explanatory virtues: e.g., simplicity, generality, conservatism, modesty, and predictive power.[3] The classic problem with this answer is that the virtues aren't obviously truth-conducive.[4] Thus Richard Fumerton:

> A theory which is more consilient and simple than alternatives [. . .] is certainly more desirable than its competitors in the sense that it would be *nice* if it turned out to be true. But this not being the best of all possible worlds (some theologians aside) what would be nice is not always so (1980, 596).

And if the virtues aren't individually truth-conducive—i.e., if it's not the case that, in general, simpler hypotheses are more likely to be true; and likewise for each of the other virtues—then it isn't clear why we should think that IBE is truth-conducive.[5]

There are a few moves we might make here. First, we might offer an independent argument for IBE's truth-conduciveness, or at least for the conclusion that we are entitled to assume as much. (See, e.g., those like Harman (1986) and Lycan (1988) who take IBE to be the fundamental form of ampliative inference, and so epistemically legitimate if any ampliative inferences are.) This maneuver would allow us to concede that, for all we know, the explanatory virtues aren't individually truth-conducive, but what matters is their collective use. Second, we might attempt to defend the truth-conduciveness of the individual virtues *as classically conceived*—i.e., as general features of hypotheses. So, for example, we might specify a sense of simplicity—number of entities, number of kinds, etc.—and then present arguments for thinking that that feature is, relative to a field of hypotheses, an indicator of truth. (Swinburne takes this line, at least with respect to simplicity. He maintains that "it is

---

[2]The adequacy requirement in the first premise ensures that h isn't merely the best of a bad lot. Granted, it matters a great deal how we articulate the relevant standard of adequacy, though I won't explore that issue here. For more on the "best of a bad lot" objection, see van Fraassen (1989) and Lipton (2004).

[3]For this list, see Quine and Ullian (1978, 64-82). See too Lycan (1988, 130) and Lipton (2004, 122).

[4]For criticisms along these lines, see Gunner (1967), Benacerraf and Putnam (1983), van Fraassen (1989), Day and Kincaid (1994), Bueno and Shalkowski (2004), and Shalkowski (2010).

[5]For more on this line of argument, see Fischer (2014).

a fundamental a priori principle" that simplicity is evidence of truth (2001, 102).[6]) Third, we might take a page from Timothy Day and Harold Kincaid, who think of an appeal to a virtue as shorthand for an appeal to a certain domain specific desideratum. On this view, we can't assess the truth-conduciveness of simplicity *simpliciter*; we can only assess the truth-conduciveness of simplicity-in-population-biology or simplicity-in-climatology. By their lights:

> IBE names an abstract pattern whose force and success depends on the specific background assumptions involved. Without substantive assumptions both about explanation in general and about specific empirical details, IBE is empty. In short, appeals to the best explanation are really implicit appeals to substantive empirical assumptions, not to some privileged form of inference. It is the substantive assumptions that do the real work (1994, 282).

So insofar as we're right to appeal to a particular virtue in a context, that's because that virtue is shorthand for a set of background assumptions that guide inferences in that context. (They make the point in terms of specifically *empirical* assumptions, but what matters is that attending to those assumptions increases the likelihood of discovering the truth, not that they be empirical *per se*.)

Let's turn to the second question that I identified above: what's the appropriate attitude to have toward the conclusion of an IBE argument?

Belief is the obvious answer, but it isn't the only one. There are various grades of acceptance that involve weaker epistemic commitments. We can see this by considering three related differences between belief and acceptance. First, belief aims at truth; acceptance need not. Instead, it can have either a pragmatic or an epistemic aim. Second, belief commits the believer to the truth of what's believed; acceptance does not. For a given purpose, we may well accept a proposition that we believe to be false. And third, belief should be based on evidence; acceptance can but needn't be. The norms governing the acceptance of propositions depend on the purpose of accepting those propositions.[7]

With these contrasts in view, one thing should be plain: if it turns out that we aren't justified in *believing* the conclusions of IBE arguments, we may still be justified in accepting them. Whether we are, of course, depends on the details of IBE, the merits of particular IBE arguments, and the relevant version of acceptance.

Of course, given some ways of construing acceptance, it isn't an epistemologically-interesting propositional attitude. For example: you may accept a proposition *p* just in case accepting *p* would improve your well-being. But on others, it certainly is. Take, for example, Greg Dawes's version of acceptance. On his view, to accept a proposition is "to employ it as a premise in one's reasoning, whether theoretical or practical, in any domain to which it might apply, with the goal of attaining knowledge" (2013, 7). The goal still involves truth—since knowledge entails it—but there is no claim to have achieved it. So we might be convinced that the best explanation is not, therefore, more likely to be true. Still, we may maintain that

[6]For other routes to simplicity's truth-conduciveness, see Quine (1963), Sober (1981), and Kelly (2007).

[7]See Tuomela (2000) for further discussion of the differences between belief and acceptance.

reasoning *from* the best explanation is more likely to lead to truth down the road. And if so, we may accept that explanation, in Dawes's sense of that term.

To sum up. Our answers to the two questions (concerning the comparative evaluation of hypotheses and the appropriate attitude toward the best one) are related. Suppose we bracket the project of giving a general defense of IBE. Then, if we can defend the truth-conduciveness of the individual virtues, we may maintain that it's permissible to believe the conclusions of IBE arguments. If we can't, then we still may be in a position to accept those conclusions, though it remains an open question whether it will be a variety of acceptance worth wanting. We need to examine the virtues to say more. So, to make progress in our understanding of theory selection in modal epistemology, we need to articulate and examine the various desiderata. Some may be applications of more general epistemic desiderata; others may be domain-specific. And once we've articulated them, we'll be in a position to begin assessing their roles in theory choice, as well as the appropriate attitude toward our selection.

4.3 The Virtues

In what follows I sketch three virtues that a modal epistemology can have: conservatism, simplicity, and the ability to handle disagreement. Granted, there are others: for example, theories are better insofar as they are internally coherent, and they are better insofar as they make successful predictions. But these are not the interesting cases: not the former, because I can't think of a modal epistemology that's been scuttled by internal contradiction; not the latter, because predictive power isn't a virtue that epistemological theories tend to enjoy.[8] Moreover, it's worth seeing how far we can get with a short list of virtues—which is, of course, the counsel of simplicity.

4.3.1 Conservatism

First, a modal epistemology is preferable insofar as it gets the cases right; it's less preferable insofar as it doesn't. That is, we want our theory to agree with our considered judgments about those modal beliefs that are and aren't justified.[9] Insofar

---

[8]We might think about getting *new* cases right as a kind of predictive accuracy. That is, for some modal claim not heretofore considered, the theory implies that we wouldn't be justified in believing it, and when we consider it, we concur that we aren't justified in believing it. But I don't see what's gained by thinking of this as predictive accuracy rather than another dimension of conservatism, so I'll stick with the latter.

[9]An example on the justified side: my coffee could be warmer than it is; an example on the unjustified side: the liquid I used to make my coffee (i.e., water) could be XYZ.

as our epistemology does this, and all other things being equal, it's preferable to one that doesn't.[10] This is an expression of conservatism.

There have been various attempts to defend conservatism's epistemic credentials.[11] By my lights, the most promising strategy is due to Quine and Ullian:

> ...at each step [opting for the conservative hypothesis] sacrifices as little as possible of the evidential support, whatever that may have been, that our overall system of beliefs has hitherto been enjoying. The truth may indeed be radically remote from our present system of beliefs, so that we may need a long series of conservative steps to attain what might have been attained in one rash leap. The longer the leap, however, the more serious an angular error in the direction. For a leap in the dark the likelihood of a happy landing is severely limited. Conservatism holds out the advantages of limited liability and a maximum of live options for each next move (1978, 67–68).

Quine and Ullian aren't claiming that more conservative hypotheses are therefore more probable than not (construing probability objectively). Rather, they are claiming that conservatism is our best long-term strategy for acquiring true beliefs—or at least for getting as close as possible. Their justification for this claim resorts to the "step vs. leap" metaphor, but it isn't hard to unpack. Suppose that we're trying to decide between two hypotheses, one of which conflicts with our current system of beliefs, the other of which dovetails nicely. If all else is equal, then it would be absurd to choose the former, since we have evidence *against* that hypothesis in the form of the evidence *for* the beliefs with which it conflicts. So, *if* we are going to take on a new belief, then, *ceteris paribus*, it should be the more conservative one.

It's important to note that conservatism, at least as understood here, doesn't provide reason to take on a new belief. (It matters that the conclusion of the last paragraph is conditional.) If we're justified in believing a hypothesis, then we have good evidence for it. An explanatory virtue is not evidence. Rather, after we've collected the evidence and identified the hypotheses that could explain it, conservatism helps us rank those hypotheses by their vulnerability to a defeater. In particular, it operates as a total evidence condition, discouraging us from ignoring the evidence we have for whatever else we believe. Hence, it discourages us from believing a hypothesis that's vulnerable to a *rebutting* defeater—i.e., evidence that supports the falsity of the hypothesis in question, thereby casting doubt on the adequacy of our original reason for believing the hypothesis. Granted, we will go wrong; we will reject true hypotheses in favor of preserving false beliefs. Nevertheless, conservatism remains a good strategy for getting closer to the truth—if only because the alternative (*not* attending to our total evidence) is certainly worse.

To see how this might play out in modal epistemology, consider Peter van Inwagen's assessment of Stephen Yablo's conceivability-based theory—namely,

---

[10]Likewise, I submit that a modal epistemology is preferable insofar as it maintains that the methods by which we acquire justified modal beliefs are the methods by which we *tend* to acquire modal beliefs. So whatever we believe about the methods by which we acquire justification about modal matters, a theory is better insofar as it spares us from revising these beliefs.

[11]See Harman (1986), Lycan (1988), Kvanvig (1989), Huemer (2001), and McCain (2008).

that it leads, with the help of a few assumptions, to modal skepticism. In short, Yablo says we are justified in believing that *p* if we can objectually imagine (imagine *de re*) a world that verifies *p*.[12] Based on several considerations, van Inwagen argues that we can't objectually imagine worlds that are significantly different from our own. So if Yablo is correct, then, according to van Inwagen, we aren't justified in believing various interesting modal claims—particularly scientific and metaphysical possibility claims. If this is right, then by the lights of anyone who thinks we *are* justified in believing those sorts of possibility claims, the *fit with the cases* criterion counts against Yablo's epistemology.

(Of course, even for someone who thinks we're justified in believing all sorts of interesting modal claims, the *fit with the cases* criterion doesn't count against Yablo decisively. Conservatism only offers clear guidance when *ceteris* is *paribus*. And that may not be the case when we consider Yablo's view. At this stage, the point is just to see how the criterion works, not to advocate for or against a particular modal epistemology.)

Conservatism already functions this way in modal epistemology, though it's rare to see it cited explicitly. Reflect for a moment on the many reactions to the problem of a posteriori necessities. (In short: if there are a posteriori necessities, then surely there are some of which we're unaware; and if there are some of which we're unaware, then we have an undercutting defeater for taking *p*'s apparent possibility to support *p*'s genuine possibility.) Most modal epistemologists don't throw up their hands at this problem. Rather, they offer (alleged) solutions. But why bother? It makes sense to offer solutions if we have some reason to avoid skepticism—for example, that we took the challenged beliefs to be justified. Otherwise, the move seems unmotivated. Correspondingly, when philosophers *embrace* skepticism, they argue as though this is a cost to be justified.[13] Why? Because all parties are sensitive to the importance of conservatism.

4.3.2 Simplicity

Hart (2003) suggests that we should posit a primitive epistemic relationship between persons and modal facts—"grasp"—roughly akin to Russellian acquaintance. Hart is forced to make this suggestion because he is staunchly committed to (a) our

[12]"Imagining can be either propositional—imagining that there is a tiger behind the curtain—or objectual—imagining the tiger itself. To be sure, in imagining the tiger, I imagine it as endowed with certain properties, such as sitting behind the curtain or preparing to leap; and I may also imagine that it has those properties. So objectual imagining has in some cases a propositional accompaniment. Still the two kinds of imagining are distinct, for only the second has alethic content-the kind that can be evaluated as true or false—and only the first has referential content—the kind that purports to depict an object" (1993, 27).

[13]On the non-skeptical side, see, e.g., Peacocke (1999), Chalmers (2002), and Williamson (2007); on the skeptical side, see van Inwagen (1998), Nozick (2003), and Hanrahan (2007).

having ample modal knowledge; (b) a neo-Quinean form of empiricism; and (c) the view that the truthmakers for modal propositions are abstracta. Whatever the merits of Hart's position, it's not simple. (The simpler conclusion is the disjunctive one: either (a), (b), or (c) is false.) And his position seems less plausible for that reason.

Simplicity is the foil of conservatism. Conservatism tells us to preserve our system of beliefs; simplicity advises us to modify it, and this to rid ourselves of the complexities we've harbored.[14] Should we trust this advice? Some balk because they think of appeals to simplicity as being predicated on a baseless conjecture about the world. Consider, for example, Paul Benacerraf and Hilary Putnam's criticism of an argument for mathematical platonism:

> ...why should the simplest and most conservative system (or rather, the system that best balances simplicity and conservatism, by our lights) have any tendency to be *true*? [...] It is hard enough to believe that the natural world is so nicely arranged that what is simplest, etc., by *our* lights is always the same as what is *true* (or, at least, *generally* the same as what is true); why should one believe that [the totality of modal truths] is so nicely arranged that there is a pre-established harmony between *our* feelings of simplicity, etc., and *truth* (1983, 34)?

Others understand simplicity as a "merely" pragmatic consideration. Thus Bas van Fraassen:

> Judgments of simplicity and explanatory power are the intuitive and natural vehicle for expressing our epistemic appraisal. [But these] are specifically human concerns, a function of our interests and pleasures, which make some theories more valuable or appealing to us than others. Values of this sort [. . .] provide reasons for using a theory, or contemplating it, whether or not we think it true, and cannot rationally guide our epistemic attitudes and decisions. For example, if it matters more to us to have one sort of question answered rather than another, that is no reason to think that a theory which answers more of the first sort of questions is more likely to be true (not even with the proviso 'everything else being equal'). It is merely a reason to prefer that theory in another respect (1980, 87).

However, simplicity need not be cashed out in a way that makes it vulnerable to these challenges. As with conservatism, the advantage of simplicity is not that it marks a particular hypothesis as probably true. Rather, it's the best way to pursue truth in the long run.

There are a number of reasons for this. First, a simpler hypothesis has a greater prior probability than its more complex cousin, at least *ceteris paribus*. The probability of p is invariably higher than is the probability of p and q; so, if we can get by without q, our hypothesis will be more probable as a result. (That is, p's posterior probability will be greater than the posterior probability of p and q—again, *ceteris paribus*. The *ceteris paribus* clause is especially important

[14]There are, of course, various ways in which one hypothesis can be simpler than another: it postulates fewer objects; it postulates fewer kinds; it is syntactically simpler; it employs fewer argument patterns to achieve the same explanatory ends (as in Kitcher's brand of unification); etc. Some ways are more important than others, but this isn't the place to explore which and why. For further discussion, see Nolan (1997).

here because you can always restate a complex hypothesis as a simple one. So we're assuming that our vocabulary has been regimented in a certain way for the purposes of hypothesis selection.) Second, simple hypotheses reduce the odds of making computational errors that might derail inquiry. This worry is not (just) about multiplying incorrectly. The more complex the theory, the greater the risk of our getting confused about the relationships between its parts, of smuggling in unstated assumptions, of misjudging how the theory should be applied in unusual contexts, etc. In minimizing these risks by preferring simplicity, we make it more likely that if our theory has to be amended or rejected, it's for the right sorts of reasons. Third, simple hypotheses leave us in a better position to handle new data. Quine makes this point nicely:

> If we encompass a set of data with a hypothesis involving the fewest possible parameters, and then are constrained by further experiment to add another parameter, we are likely to view the emendation not as a refutation of the first result but as a confirmation plus a refinement; but if we have an extra parameter in the first hypothesis and are constrained by further experiment to alter it, we view the emendation as a refutation and revision. Here again the simpler hypothesis, the one with fewer parameters, is initially the more probable simply because a wider range of possible subsequent findings is classified as favorable to it (1963, 105).

In this last respect, simplicity amounts to a rejection of Popper's call to make bold conjectures. But that's as it should be. Bold conjectures aren't well-supported by evidence; they're designed to be falsified. When we're trying to decide which hypothesis to believe, we want to avoid risk, not embrace it.

Finally—and perhaps in light of the above—we can see "prefer the simple" as a fast and frugal heuristic that guards against ad hoc hypotheses. We don't need justification to *abstain* from saying that the world is a certain way; rather, we need justification to make such claims. So if we can explain the data with a weaker theory, and we take it on a stronger one anyway, we should worry that we've done so for the wrong reasons—namely, preserving a pet theory. Thus, "prefer the simple" checks the temptation to resist theoretical upheaval. This isn't evidence that, in a field of competitors, the simplest hypothesis is more probable than not. But it does suggest that, if we want to reach the truth, it would be foolish not take the road paved with simple hypotheses.[15]

In the case of Hart's "grasp," we might not immediately see its problems in the terms outlined above. Instead, we might want to express our reservations with a series of questions: Do we ever grasp the wrong thing, or only partially grasp the right thing? If so, what explains this? Can we detect such errors? If we can't, then what's the epistemic significance of this fact? Furthermore, how did grasp evolve? What advantages did it confer on our ancestors? Is there reason to suppose that it's a uniquely human trait, or should we expect other species to have it as well? I think these questions illustrate the costs of increasing qualitative complexity—in this case, postulating a new kind of relation. Hart's view is bound to involve features

[15]For a formal argument for the same conclusion, see Kelly (2007).

that we find it difficult to deploy; it is a bold conjecture (and yet one not obviously falsifiable); it appears to be ad hoc. So, simplicity counsels against it.

For the same reason, simplicity may also counsel against accepting views that involve a fortuitous correlation between our modal beliefs and modal reality. For example, Scott Sturgeon argues that modal rationalism—like that of Chalmers (2002)—is vulnerable to "the Objection from Magic." That is, such accounts posit a perfect and inexplicable correlation between primary, positive, ideal conceivability and metaphysical possibility (see Sturgeon (2010); cf. Kallestrup (2006)).[16] If Sturgeon is right, then modal rationalism might not be much of an advance over "grasp." Similarly, simplicity might counsel against the theistic reliabilism that O'Connor (2008) develops based on Plantinga (1993). According to O'Connor, God ensures that our modalizing produces true beliefs (when performed properly in the right environment). This view can't be faulted for positing an *inexplicable* correlation between our modal beliefs and modal reality—God stands in the gap—but insofar as philosophers regard theism itself as an unduly complex hypothesis, we might not think that simplicity is on O'Connor's side.[17]

On one level, all these examples highlight the difficulty of handling "the integration challenge"—the term coined in Peacocke (1999) for the task of reconciling our epistemological and metaphysical convictions. This challenge is especially difficult if we deny that we're causally related to the truthmakers for modal claims (because we take them to be abstracta, or Lewis's concrete worlds, etc.) yet remain optimistic about the scope of our modal knowledge. It wouldn't be surprising, then, if simplicity tended to push us toward various ways of reducing modality, thereby dissolving the problems that animate the modal epistemology literature. But insofar as we resist the reductive move, I suspect that simplicity recommends either (a) greater epistemic humility or (b) beginning with views that extend the epistemic resources we use in other domains—for example, induction, abduction, analogical arguments, etc.—rather than postulating new ones.[18] (Of course, it might also recommend both, which would be convenient for TEM.)

[16]Of course, we might also doubt that the correlation obtains; on this point, see Vaidya (2008).

[17]On the other hand, God makes available a kind of Augustinian platonism, which means that you don't have to postulate independent abstract objects, and O'Connor uses God's power as what grounds modal facts. Simplicity might *favor* O'Connor's view if you can explain all this by postulating a single being.

[18]See, e.g., Roca-Royes (2007), Leon (2009), and Biggs (2010). On a different note, someone might worry that we need to be able to assess the modal consequences of competing theories in order to assess their relative conservatism or simplicity, which requires having modal justification to get modal justification. So is it circular to rely on the virtues? No. Recall Sect. 3.3: conditional modal knowledge is cheap, and that's the only information that's required at the comparison stage.

4.3.3 Disagreement

We know how to adjudicate perceptual disagreements. If I say that my keys are on the table, and my wife says that I left them on the dresser in the bedroom, then it doesn't take much to figure out how to settle our dispute. It would be nice to have a modal epistemology that made it equally easy to resolve disagreement, as unrealistic as that might be. Still, it's no doubt a virtue of modal epistemology if it makes it *easier* to sort out differences of modal opinion—and the more tools if offers, the better.

Handling disagreement is not a standard explanatory virtue. In the present context, however, it deserves special mention.[19] First, we might argue for its importance by maintaining, with Yablo (1993), that this helps to protect against *brute* disagreement, which can be evidence that the disagreement isn't about an objective fact (in some sense of that vexed phrase; Yablo's line on this draws on Wright (1986)). After all, if our modal beliefs aren't about modal facts, then our work in modal epistemology is almost done; after we develop our error theory, we can go home.[20] Thus handling disagreement is a way to avoid a potential defeater: namely, that we have reason to think that there are no modal facts, and thus reason to doubt that our modal beliefs can be either true or false.

But even if we part ways with Yablo, maintaining that brute disagreement does *not* count against the objectivity of the fact in question, it's surely the case that a brute disagreement about p can undermine our justification for believing p. (We can doubt our methods for acquiring beliefs about x without doubting that x is real.) Again, this is not to present handling disagreement as a mark of truth, but only as a way to avoid a criticism. Insofar as we take ourselves to have justified beliefs about certain modal matters, and insofar as those matters are contentious, we have cause to seek a modal epistemology that will help us to diagnose the controversy. Otherwise, it may be difficult to avoid the conclusion that the disagreement is brute, and thus that our justification was illusory. This isn't just an expression of conservatism, though it's that as well. Additionally, it's a way of replying to the charge that *if* we've struck the truth, it's only by chance. If we can explain where our peers go wrong, then we're in a better position to explain what it takes to go right, and thus why we might be reasonable in trusting our judgments about what is and isn't possible.

There are plenty of proposals about the nature of modal error, which can—in general—be reframed as accounts of how to negotiate modal disagreements: for example, Yablo (1993, 2006), Chalmers (2002), Bealer (2004), Gregory (2010), and Sturgeon (2010). But my own view is that, insofar as any account is promising, it's because it draws our attention to the larger doxastic context in which modal judgment occurs.

---

[19]There may well be a way to construe this virtue in terms of conservatism or simplicity, but in the present context, little would be gained from trying.

[20]There are alternate accounts of modality on which this isn't true; see, e.g., Thomasson (2007a, b).

To see what I mean, let's begin with the observation that there is something appealing about the perceptual metaphor we sometimes employ when discussing certain modal phenomena. When I contemplate red and blue, I just 'see' that they are incompatible, that it is *impossible* for any wholly red surface to be wholly blue. But we should admit that the story has to be far more complicated than the metaphor suggests. Consider a *thermochromic* surface—that is, a surface composed of a substance such that its color varies depending on its temperature. (Liquid crystals, the substances used to make many thermometers and mood rings, are prime examples.) There certainly are thermochromic surfaces that can be wholly red at one time and wholly blue at another. Is this a counterexample to the impossibility claim above? Perhaps not. Perhaps this example only shows that the original formulation of the claim was sloppy, and the correct thing to say is that no surface can be wholly red and wholly blue *simultaneously*. But now we're faced with a hard question: on what basis do we deny that the surface is wholly red and wholly blue simultaneously? Why, for example, do we think that the relationship between *being wholly red* and *being wholly blue* is not akin to that between *being fragile* and *being resonant*? Those properties may not be manifested simultaneously either, but there is no obvious reason to deny that, say, a wineglass has them simultaneously. The point here is not that we should adopt some quirky version of color dispositionalism. Rather, the point is that, when we say we "see" that being wholly red is incompatible with simultaneously being wholly blue, we're assuming—*inter alia*, and probably without realizing it—that being a particular color is *not* like a thing's having a dispositional property.[21] (This, I think, is what Lowe (2008) gets right when he argues that modal information informs our perceptual judgments. He's just wrong that we can only justify the relevant claims via *a priori* methods.)

Such assumptions often sit in the background when we modalize; it's rare that we make them explicit and put them on the table for consideration. They also tend to be the differences that make disagreements so hard to resolve. Consider, for example, the various Kripke-style arguments for dualism.[22] Such arguments rely on certain modal premises, all of which have faced sustained criticism, though this fact seems not to demoralize their advocates. As far as I can tell, this is because one philosopher's view of *x* might be so different from another's that the first sees no reason to think that F*x* is possible, while it seems obvious to the second that it is.

[21] Someone might object that this isn't an assumption—it's a conclusion, and one that's warranted based on the seeing. I find this implausible. First, if it's a conclusion, then it appears to be one that's reached without considering alternatives; and, if that's right, then one wonders why we should take it to be warranted. Second, note that we have a fairly clear account of how we can "perceive" sophisticated facts thanks to relevant prior knowledge: the neurologist "sees" frontal cortical atrophy on a CT scan, whereas I see a Rorschach test, because of what the neurologist knows about brains, CT scans, and the like. And absent such knowledge, the perception would be completely mysterious. So, barring considerations to the contrary, we should offer a model of our perception of modal facts that's in line with our model of the perception of actuality. (Of course, in the modal case, the assumption is no doubt implicit, but that makes no difference: what matters is that background information constrains the interpretation of the data.)

[22] See, e.g., Kripke (1980), Taliaferro (1986), Hart (1988), Yablo (1990), and Chalmers (1996).

For example, the first thinks that pain is characterized by its feel; the second that it's characterized by its functional role in an organism.[23] These sorts of standoffs have made us wary of such modal arguments, since they appear to circumvent, rather than face squarely, the substantive issues: namely, what's the right theory of pain? Progress with respect to modal disagreement, then, may involve shifting our attention to background theories, implicit assumptions, and their ilk. A modal epistemology can aid us in making this shift by drawing attention to the ways in which our justification is mediated, at least in part, by their doxastic context.

Again, an epistemology's ability to handle disagreement is not a mark of its truth. Nevertheless, like conservatism and simplicity, it's a valuable constraint in our search for the truth. For without it, we might well adopt a theory that lacks the resources to answer two allegations that should worry anyone who takes modal knowledge to be a worthy subject of inquiry: first, that there are no modal facts; second, that we aren't justified in believing any controversial modal claims.

4.4 Belief or Acceptance?

Before going further, let's recall the formulation of IBE with which we began:

1. Hypothesis *h* would, if true, adequately explain facts f_1—f_n.
2. *h* would, if true, explain f_1—f_n better than any of its competitors.
3. So, probably, *h* is true.

This formulation raises two questions. First, in what sense is *h* the best explanation? Second, what's the appropriate attitude toward the conclusion of an IBE argument? In the last section, I offered three virtues that a modal epistemology can have: conservatism, simplicity, and the ability to handle disagreement. On my view, then, if *h* is a modal epistemology, it qualifies as the best explanation in virtue of earning the best overall score on these three virtues. I can't precisely unpack the 'score' metaphor I just deployed. Nevertheless, I take that we are somewhat familiar with making such comparative evaluations in practice.

This still leaves the problem of balancing three desiderata. We can make some progress on this issue by reflecting on the virtues we've discussed. Conservatism, recall, is akin to a total evidence condition; it guards against adopting hypotheses that face rebutting defeaters. So, we can think of it as having a regulatory function, to be deployed *after* the other virtues, not before. (Deploying it first would be to risk dogmatism; deploying it simultaneously creates the problem rather than solving it.) This two-stage approach makes our task somewhat easier. Now, our aim can be to look for simple hypotheses that can handle disagreement. Once we have a field of competitors meeting this description, we can assess them in terms of conservatism.

[23]This is, essentially, the criticism leveled in Jackson et al. (1982).

This is hardly a decision procedure, but it's nevertheless an advance over a single-stage process, which provides no real advice about how to adjudicate the demands of competing desiderata.

This leaves us with our question about the appropriate attitude. In the preceding section, I was careful in the way that I articulated the epistemic significance of the virtues. By my lights, being the simplest hypothesis in a field of competitors is not evidence that the hypothesis is in fact true. So, I grant that we shouldn't believe a hypothesis on the basis of its simplicity. Still, when choosing between competing hypotheses, we have every reason to prefer the simplest—at least if *ceteris* is *paribus*—because our aim of securing the truth in the long run is better served thereby. I say the same of the other virtues. Therefore, we needn't settle for an epistemologically-uninteresting attitude toward the conclusions of IBE arguments in modal epistemology; the discussion warrants more than using-for-the-sake-of-convenience. Instead, we can recommend in modal epistemology what Dawes recommends generally: if a hypothesis fares the best on the explanatory virtues, then we should accept it—*not* believe it. Nevertheless, we haven't crossed over into the merely pragmatic. Truth, after all, is the aim: to accept a proposition is "to employ it as a premise in one's reasoning, whether theoretical or practical, in any domain to which it might apply, with the goal of attaining knowledge" (2013, 7). Dawesian acceptance is a properly epistemic propositional attitude, and so we can reformulate IBE as follows:

1. Hypothesis h would, if true, adequately explain facts f_1—f_n.
2. h would, if true, explain f_1—f_n better than any of its competitors.
3. So, we should accept h.

This is a weaker version of IBE, and it's a better one for that reason. Granted, it isn't clear why we should postulate a tight connection between the explanatory value of a particular hypothesis and the truth of that hypothesis. However, this version of IBE dodges such worries, since it's clear why there's a link between explanatory value and acceptance, as well as how acceptance—at least as Dawes construes it—is linked to truth in the long run.[24]

[24]That said, I don't mean to imply that the case for acceptance is watertight. Our judgment to accept, rather than believe, is based on our threshold for epistemic risk. Our threshold of epistemic risk is the point at which we are unwilling to venture belief based on the evidence; it's the point at which we judge that the risk of believing falsely outweighs the possibility of believing truth. If we put thresholds on a simple spectrum, then it would be one that runs from skeptic to sucker. Skeptics have a very low threshold for epistemic risk, and hence they deny that we know much of anything. Suckers, by contrast, have a high threshold; they are eternal optimists when it comes to the scope of their knowledge. Let's grant that we shouldn't set our threshold at either extreme. Might there be a range of acceptable thresholds in the middle? Perhaps so. And in the present case, I see no obvious reason to deny that, with the right argument behind it, tentative belief could be an appropriate attitude. For more on ranges of epistemic risk, see Riggs (2008).

4.5 IBE and TEM

We now have some sense of the criteria for choosing between modal epistemologies, how to employ them, and the attitude they can warrant. How does TEM fare? Let's consider the criteria in reverse order.

4.5.1 Disagreement

First, TEM is very useful for understanding (and sometimes resolving) modal disagreements. This is because TEM gives us a battery of questions to ask when faced with competing modal claims. Initially, we should inquire as to whether the two parties are reasoning from the same theory. If so, then is it actually clear that the theory speaks to this issue? Is the theory inconsistent? Are the models in question clearly supposed to represent the target system? Has someone made a computational mistake? Suppose that the two parties hold different theories. In that case, after confirming that both theories make a pronouncement about the topic at hand, we should begin to consider their respective merits. What phenomena do the explanations purport to explain? Are they offered to explain the same data? If not, then is there any basis for preferring one set of data to the other? And if they are offered to explain the same data, then how well does each explanation score on the virtues? Can you show that one is superior to the other, and hence that the modal claims following from it are more worthy of belief? If you can't, then is this a case of reasonable disagreement, or should the dispute undermine the confidence of both parties? No doubt there are still more questions that we could ask, but the ones mentioned are illustrative of how we can use TEM to locate the source of a modal disagreement—and the prospects for resolving it.

Plainly, TEM's approach to modal disagreement is based on there being a tight relationship between the justification of a modal belief and the justification of the corresponding theory. Most modal epistemologies treat the justification of our modal beliefs as largely independent of the justification of our non-modal beliefs.[25] And if you think that our modal and non-modal beliefs are justified in (more or less) independent ways, then you'll have no trouble with the notion that there can be arguments in which all the disputants agree about the non-modal facts while disagreeing about the modal facts. If TEM is true, though, then this sort of case is very unlikely, if it's possible at all. This is as it should be, since it is very plausible that the modal supervenes on the non-modal—i.e., there is no difference

---

[25] It's interesting to consider whether identities are an exception to this generalization. However, you might well deny that identity claims are non-modal—at least if by "identity" we mean *identity*, and not just a relation that's actually symmetric, transitive, and reflexive.

in the modal facts without a difference in the non-modal facts.[26] TEM directs us to refocus our attention on the non-modal facts, since there very likely *isn't* genuine agreement about them, and that disagreement is likely to provide some insights as to why the disputants are at loggerheads over this or that modal claim. After all, the corresponding non-modal claims will be about objects and properties that were posited for theoretical purposes, and even slight disagreements about the relevant explananda, or about what counts as the best explanation thereof, may lead to radical differences down the line.

I think that modal disagreement is perhaps the area where TEM can be most fruitful, although I grant that much more should be said about it. Still, the above should be enough to indicate how TEM might be deployed in cases of modal disagreement, and the tools that it offers certainly seem promising.

4.5.2 Simplicity

TEM is also a simple theory. If you think that we justifiably believe our best theories, and if you are sympathetic to the claim that theories have modal implications, then you have the building blocks for TEM. These are not wildly complex views, and TEM does not demand much more of us than that we conjoin them in a particular way. Therefore, it does not require us to multiply sources of justification; it doesn't even require us to extend or retool epistemological principles to which we are already committed. TEM simply points out that a plausible account of theories and their justification *includes* a modal epistemology: if our scientific theories are justified, and if we allow their content to be represented by sets of models, then we have essential components for TEM. The only additional claim—or, rather, the only essential additional claim—is that this route to interesting modal knowledge is the only one available.

Contrast TEM with Hart's grasp, or Chalmer's perfect and inexplicable correlation between primary, positive, ideal conceivability and metaphysical possibility, or O'Connor's theistic reliablism. Of course, these views aren't the only options in the field, but they highlight what TEM *doesn't* need to postulate, and its relative simplicity for that reason.

---

[26] I take it that, for every (purportedly) non-modal facts, some modal fact or other supervenes on it. When a supervenience relation is universal in this way, one begins to wonder about the plausibility of the claim that the supervenience base and the supervening facts are truly distinct. And insofar as a claim's being non-modal depends on its subject matter being non-modal, the supervenience of the modal on the non-modal provides reason to deny that there are truly non-modal claims.

4.5.3 Conservatism

TEM provides tools with which to handle disagreement; TEM is simple compared to some of its competitors. Is it conservative?

By some metrics, the answer is Yes. Any scientific realist will affirm that we justifiably believe at least some of our best theories, and scientific realism is surely the dominant view amongst philosophers and non-philosophers alike. Though the semantic view has its detractors, and while there is disagreement about how its details should be worked out, it still appears to be the new orthodoxy amongst philosophers of science. So TEM's core commitments are conservative.

Moreover, the call to inspect our theoretical commitments is conservative. When someone maintains that *p* is possible, and *p* is contentious, we naturally wonder about the theses to which that person is thereby committed. If Jones says, for example, that he could be disembodied, we would be right to ask what he believes in virtue of which this seems possible to him. Does he believe in Cartesian souls? Aristotle's substantial forms? Does he think that the mind is to the brain as software is to hardware? (If so, what runs the software when he's in a disembodied state?) Does he think that he's an abstract object? A set of time-slices, perhaps? (If a set, then isn't it true that, strictly speaking, he is *necessarily* disembodied?) Suppose that Jones denies that he believes any of these things, and he's unwilling to offer an alternative hypothesis. What, then, should we make of his claim that he could be disembodied? I suspect that many of us will doubt that he's entitled to it; we'll doubt that he has really thought the matter through. To my mind, this is some evidence that we are looking for a theory behind Jones' modal claim—for some hypothesis about the world that lends it credibility. When we fail to find one, we stop taking his claim as seriously.

Additionally, TEM is conservative with respect to how it encourages us to resolve modal disagreements. Suppose we propose a modal claim, citing our ability to imagine that *p* as some evidence for *p*'s possibilty. Jones resists: "Of course I can *imagine* that *p*," he says, "but so what?" If we want to persuade Jones, we can do no better than to show him that *p*'s possibility follows from a theory that he accepts. If Jones initially rejects *p*'s possibility, why does he defer to the theory later on? One appealing explanation is that we're *always* willing to defer to some theory or other, but because many of our theoretical commitments are shared, and because we rarely consider weird modal claims, it usually suffices to appeal to the imagination, the deliverances of which are shaped by the theories that we accept. When confronted with more unusual claims, we aren't sure whether to trust our imaginations; we want some confirmation that they are not leading us astray. Hence, we make explicit recourse to our theories in those contexts. Insofar as this explanation is plausible, it lends some credibility to the claim that TEM harmonizes with our antecedently-held beliefs about how to adjudicate disputes about modal matters.

It's also worth noting that TEM is conservative in that it fits with a plausible hypothesis about the relationship between knowledge of actuality and knowledge of modality. It would, for example, be strange to be a complete skeptic about our knowledge of this world while being wildly optimistic about our knowledge of other worlds. At the outset of inquiry, the working assumption should be that our views about the potential scope of our modal knowledge will be correlated with our views about the potential scope of our knowledge of actuality.[27] If TEM is true, this is so. If you think, for example, that we are really only justified in believing what our best science tells us, then TEM explains why it's appropriate for you to take a similarly high view of the modal claims that our best science underwrites *and* a correspondingly low view of modal claims that follow from non-scientific theories (e.g., folk and metaphysical theories). But if you think, on the other hand, that some of our metaphysical theories are justified, then TEM explains why you should take a more generous stance on the potential scope of our modal knowledge, allowing for belief about the physically impossible but metaphysically possible (on the assumption that your metaphysical theory does not entail that what's physically necessary is necessary *simpliciter*). So TEM fits with a plausible thought about how two epistemic domains are related.

Lastly, TEM is conservative insofar as it fits with the epistemic principle that emerges through the game analogy, introduced earlier: when we consider interesting modal claims with a game, we should rely on our theory of gameplay; likewise, when we consider interesting modal claims about the world, we should rely on our best theories about the world. Again: the modal facts aren't ours to stipulate, and guesses will be guesses—not justified beliefs. So, we have to try to extrapolate the facts in question from the only data available, which are our observations of the world. (A point that, of course, doesn't preclude bootstrapping.)

All that said, we should acknowledge that many won't be inclined to see TEM as conservative in one crucial respect. Typically, we say that an epistemological theory is conservative if it gets the cases right, which is to say that it agrees with our pre-theoretic judgments about what we are and aren't justified in believing. Many philosophers are inclined to think that we're justified in believing all sorts of claims—e.g., that bars of iron could float in water, or that the gravitational constant could have been something other than what it actually is—that TEM says we aren't justified in believing. This isn't because TEM says that we *couldn't* be justified in believing them. Rather, it's because TEM says that we're only justified

[27] I don't know whether Cartesian skepticism is the exception to my claim that there is a natural correlation between (alleged) knowledge of actuality and (alleged) knowledge of possibility and necessity. The skeptic denies that we justifiably believe anything of which we are not certain. Presumably, this rules out ordinary and scientific theories but rules in some logic, and therefore the skeptic is likely to regard consistency as a guide to possibility. At first blush, then, we should say that the skeptic is entitled to the modal premises that she employs in her skeptical arguments, since—for example—it does seem that the evil demon scenario is consistent. If this is right, then skepticism is the exception to my rule. However, I wonder whether the skeptic should be *certain* that the evil demon scenario is consistent. What would it take to show this? I am not sure, and I doubt that she is either. This doesn't create a problem for skepticism—since it may suffice for her purposes if the evil demon scenario is merely epistemically possible—but it is enough to keep my correlation principle intact.

in believing them if we justifiably believe theories according to which they're true. And presumably, we don't. In any case, to distinguish this feature of TEM from others relating to conservatism, let's say that TEM is *cautious*.

I can bite the bullet here: I can say that while TEM has many virtues, it just isn't cautious, and some will reasonably judge that it's worse for that reason. However, I don't think I need to go that route. First, based on the game analogy, I think that caution is appropriate. Second, I deny that TEM's cautiousness is a cost relative to its competitors, since I think that they're more cautious than people often take them to be. Or, at least, so I'll argue in Chap. 6. So, I want to argue both that TEM's cautiousness is good in itself, and that if you disagree, it still isn't worse than others for this reason. To help make the former case, it will help to get clearer on TEM's cautiousness. Perhaps the best way to do that is to clarify the relationship between TEM and modal skepticism. As we'll see in Chap. 5, TEM provides quite a different basis for modal skepticism than the one van Inwagen develops, and one that leads to a more principled modesty about modal matters.

Chapter 5
TEM and Modal Skepticism

Abstract In this chapter, I argue that TEM is offers a better basis for modal skepticism than the one van Inwagen develops. My aim here is twofold. On the one hand, I want to show that TEM *can* lead to a form of modal skepticism, and if it does, then it leads to a better form than what's previously been available. Second, I want to clarify what it would take to get a non-skeptical version of TEM—i.e., one on which we're justified in believing some extraordinary modal claims. In other words, the goal here is to give a better sense of what TEM requires of those who think, for example, that they justifiably believe that there could be a maximally perfect being. Doing so will help us appreciate the nature of TEM's cautiousness, as well as what's valuable about it.

What does TEM suggest about our epistemic limits, both current and absolute? Very roughly, TEM says that, where *p* is any interesting modal claim, we're justified in believing *p* only if we are justified in believing a theory that says that *p* is true. It follows that the *current* potential scope of our modal knowledge is determined by our current theories: if they pronounce on *p*'s modal status, then we could justifiably believe that it's true or that it's false; if not, then we couldn't justifiably believe either.[1] The *absolute* potential scope of our modal knowledge is determined by those theories that we could come to justifiably believe: if we could come to justifiably believe a theory that pronounces on *p*'s modal status, then we could justifiably believe that *p* has the relevant status; if not, then we couldn't.

What is van Inwagen's modal skepticism? To answer this question more precisely, let's ignore technological and scientific modal claims—which are among the interesting ones—and move beyond the rough and ready characterization of extraordinary modal claims that we've been using. According to van Inwagen, a proposition—not a modal claim specifically—is *extraordinary* if it meets these conditions: (1) it concerns matters "remote from the practical business of everyday life," (2) its truth-value "cannot be determined by logic and reflection on the meanings of words or by the application of mathematical reasoning," and (3) its truth-value is "unknown to us or... known to be false" (1998, 74, 84 n. 17).

[1] I am suppressing concerns about conflicts between theories and undercutting defeaters (like Benacerraf's Dilemma).

B. Fischer, *Modal Justification via Theories*, Synthese Library 380,
DOI 10.1007/978-3-319-49127-1_5

His modal skepticism is the view that, once we attach modal operators to such propositions, we don't and can't know their truth values.[2] Van Inwagen thinks that this view rules out knowing modal claims like these:

- Necessarily, there exists a being that has all perfections essentially.
- Possibly, I exist and nothing material exists.
- Possibly, there exists vast amounts of suffering for which there is no explanation (1998, 67–68).[3]

So modal skepticism is *not* the view that we don't have any modal knowledge whatever. Rather, it's modal *modesty*—and in particular, modesty about many of the modal claims that professional philosophers are happy to deploy in their arguments.

In this chapter, I argue that TEM is offers a better basis for modal skepticism than the one van Inwagen develops. My aim here is twofold. On the one hand, I want to show that TEM *can* lead to a form of modal skepticism, and if it does, then it leads to a better form than what's previously been available. Second, I want to clarify what it would take to get a non-skeptical version of TEM—i.e., one on which we're justified in believing some extraordinary modal claims. In other words, the goal here is to give a better sense of what TEM requires of those who think, for example, that they justifiably believe that there could be a maximally perfect being. Doing so will help us appreciate the nature of TEM's cautiousness, as well as what's valuable about it.

Here's the plan. First, I set out van Inwagen's case for modal skepticism. Most of its elements are present in his (1998), but he has long been advocating for a more cautious theory of modal knowledge, so I will be drawing from some of his

[2]I'll note here that TEM-based modal skepticism won't be equivalent to van Inwagen's modal skepticism without some significant assumptions. Van Inwagen's view involves agnosticism about both extraordinary possibilities and necessities. You can make a TEM-based version of modal skepticism say the same thing. However, it requires assumptions to the effect that (a) metaphysical possibility outstrips physical possibility and (b) we ought to interpret the laws of physical theories as making claims merely about what's physically necessary. Absent these assumptions, TEM is actually less cautious, and this for reasons familiar from Chap. 3, Sect. 3.1. As I argued there, unless you're justified in believing a theory according to which the physical possibilities are a proper subset of the metaphysical possibilities, you've got no reason to think that there are any physically-impossible metaphysical possibilities. Hence, you should trust your best theories to guide your judgments about modal reality, however tentative that trust might be.

[3]Note that van Inwagen does *not* think that a proposition is extraordinary just in case it is philosophically contentious. There are, for example, those who deny that all identity claims are necessary if true, but he's open to the view that we know the detractors to be in error (1998, 82, n. 11). Also, he admits that the view's name "was perhaps ill-chosen, since, as I have said, I think that we do know a lot of modal propositions, and in these post-Cartesian days, 'skeptic' suggests someone who contends that we know nothing or almost nothing. It should be remembered, however, that there has been another sort of skeptic: someone who contends that the world contains a great deal of institutionalized pretense to knowledge of remote matters concerning which knowledge is in fact not possible. (Montaigne was a skeptic in this sense, as were, perhaps, Sextus and Cicero.) It is in this sense of the word that I am a modal 'skeptic'" (1998, 69).

earlier papers too. Then, I consider a few problems with van Inwagen's argument. After offering the best solution I can, I contend that van Inwagen's variety of modal skepticism rests on an untenable analogy between modal and perceptual knowledge. Next, I show how you could use TEM to dodge these problems and develop a more plausible version of the view. With these foundations in place, I discuss what it would take to generate a version of TEM that didn't lead to modal skepticism. As we'll see, it's hard to do, but that challenge is the right one to have.

5.1 Van Inwagen's Argument

I understand van Inwagen's overarching argument as follows.[4] If you believe that *p* is possible, then you are committed to there being a possible world at which *p* is true. If there is such a world, then there are various historical and nomic facts with which *p*'s truth is compossible. So, if you believe that *p* is possible, you are committed to there being a history and set of laws with which *p*'s truth is compossible. Suppose you recognize this (as you should, if you are a philosopher). Then, you justifiably believe that *p* is possible only if you justifiably believe that there is a history and set of laws with which *p*'s truth is compossible. Now let's suppose that *p* is an extraordinary proposition (i.e., one (1) that concerns matters "remote from the practical business of everyday life," (2) whose truth-value "cannot be determined by logic and reflection on the meanings of words or by the application of mathematical reasoning," and (3) whose truth-value is "unknown to us or [is] known to be false." You should be agnostic about whether there is a history and set of laws with which *p*'s truth is compossible. Hence, you should be agnostic about *p*'s possibility.

Let's say that there is a "*p*-friendly world" just in case there is a history and set of laws with which *p*'s truth is compossible. The key claim in this argument is that you should be agnostic about whether there is a *p*-friendly world. Why believe it?

There are a number of pieces to van Inwagen's answer, but the most important one is that possible worlds are *coherent* entities—their parts "hang together" in some important sense of that phrase. He puts this as follows:

> To assert that *p* is possible [...] is to commit oneself, willy-nilly, to the thesis that there is a whole, coherent reality—a possible *world*—in which *p* is true, of which the truth of *p* is an integral part. To assert that it is possible for the moon (or a thing in the moon's actual orbit that looks like the actual moon when observed from the surface of the earth) to be made of green cheese is to commit oneself, willy-nilly, to the thesis that a physical universe in which a moon-like thing made of cheese came into existence and continues to exist is possible, that there are possible laws of nature and possible initial conditions that permit such a thing. (Or, if the object is supposed not to have arisen in the natural course of events,

[4] As I said, I'm extracting this line of reasoning from van Inwagen (1998). However, he defends a similar line in a number of other places. See, e.g.: van Inwagen (1977, 1979, 1997, 2006).

> but to be literally miraculous, to the thesis that there could be a supernatural agency that was capable of creating and sustaining it and which either had a good reason to create and sustain a huge ball of cheese in orbit around the earth, or which might create and sustain such a thing without having any good reason to do so.) (1998, 77–78)[5]

Given this, then it becomes much, much harder to see how we could come to know that an extraordinary proposition is possibly true. Take, for example, the suggestion that we can know that the moon could be made of green cheese by imagining this state of affairs:

> ...anyone who thinks he can imagine that the moon is made of cheese has a very sluggish imagination: the active imagination demands a pasture for the antecedently necessary thousands of thousands of millons of cows, demands a way to preserve a piece of cheese in broiling heat, freezing cold, and vacuum for thousands of millions of years, demands some off-stage *machina* to protect a piece of cheese thousands of miles across from gravitational compression into non-cheese, demands... But any *serious* attempt to imagine the moon being made of green cheese—and what besides a serious attempt could prove "coherence"?—must, like the unimaginable object itself, soon collapse under its own weight (1979, 671–672).

The upshot is this: van Inwagen thinks that we're justified in believing that p is possible if we can provide good reason to think that there is a world p-friendly world, but he despairs of our ability to do this for extraordinary propositions.

There are two natural reactions at this juncture. First, you can doubt whether the task before us is as difficult as van Inwagen suggests: is it *really* that hard to imagine a p-friendly world? Second, you can grant that it's hard to imagine a p-friendly world, but doubt that we never succeed: is it plausible that our imaginings are *never* sufficiently detailed to provide evidence that a p-friendly world exists?

Van Inwagen thinks that a confused idea stands behind the first reaction: namely, that consistency is a reliable test for possibility. He contends that logical

---

[5] I don't think that van Inwagen regards all this as contentious, even though the passage may commit him to denying various Humean theses—e.g., that the laws are just the axioms of the true deductive system that best balances simplicity and strength, that there are no necessary connections between distinct existences, that (some version of) the principle of sufficient reason is false. But since (a) I agree with everything that van Inwagen says in this passage and (b) I don't have space to explore the merits of Humeanism, I'd like not to pursue this issue further. Perhaps it will suffice to say the following. First, if it turns out that van Inwagen's argument turns on various non-Humean claims, then all the better for me: I regard it as a virtue of TEM that it ties our modal knowledge to our best theories, and I am glad to take the relevant non-Humean claims to be (part of) a theory. This shifts the debate to the metaphysical theories themselves, and if nothing can be said in their favor that doesn't presuppose contentious modal claims, then so be it: we will have to accept silence about the relevant modal matters. Second, it is hard to provide a non-circular argument for *any* metaphysical thesis, and insofar as our metaphysical beliefs are justified, it seems to me that that justification is quite weak. (And again, it's worth pointing out that many arguments for large-scale metaphysical views—like Humeanism—are bound to rely on modal claims that are in question here.) So, an epistemology of modality should be very cautious about making metaphysical assumptions—including Humean ones. This point also cuts against van Inwagen, of course, but if it's true, then it's enough to get the essence of his conclusion: since we should be not be so confident that the metaphysical facts are congenial to our epistemic practices, we should not be so quick to say that p is possible, at least where p is an interesting proposition.

"possibility" is not a species of absolute possibility (i.e., metaphysical possibility or possibility *simpliciter*). We know this because we know that there are necessarily false propositions that are not contradictory: e.g., some things are simultaneously red all over and green all over, there are male vixens, and I am a donut.[6] It follows that the space of logical "possibilities" includes some absolute *im*possibilities. (Contrast this with the space of physical possibilities—those absolute possibilities consistent with the actual laws of nature. Not all absolute possibilities are physical possibilities, but every physical possibility is absolutely possible; hence, physical possibility is a species of absolute possibility.) Now, if a proposition does not appear to be contradictory, then you are probably entitled to infer that it is not logically impossible. But this is to make an epistemological observation, namely, that the proposition in question cannot be shown to be absolutely impossible by logic alone. If logical possibility *were* a species of absolute possibility, then you could infer from its (probably) not being logically *im*possible to its (probably) being absolutely possible. But since it isn't, you can't. (*Objection:* There are very few absolute impossibilities that aren't logical impossibilities; hence, logical possibility is an imperfect but nevertheless genuine guide to absolute possibility. In other words, while it's true that the space of logical possibility is larger than the space of absolute possibility, the former isn't *that much* larger than the latter. *Reply:* Both "spaces" contain indenumerably many worlds; finding out that one isn't *that much* larger is cold comfort. Moreover, if you know that there are very few absolute impossibilities that aren't logical impossibilities, then you must know what is and is not absolutely impossible; otherwise, you'd be in no position to judge the ratio of absolute to logical impossibilities. How did you come by this knowledge?) The upshot: you don't know that there is a history and set of laws with which p's truth is compossible—i.e., you don't know that there is a p-friendly world—just because you detect no contradiction when you search for one.

The second reaction is to doubt that our imaginings are *never* sufficiently detailed to provide evidence that a p-friendly world exists. But van Inwagen simply denies that we have the resources necessary to assess extraordinary claims: in some cases, the task is just too complex for us; in others, it requires information that we don't have. He uses Stephen Yablo's (1993) conceivability-based modal epistemology to illustrate this point. (He seems to regard Yablo's view as correct in outline, if not in detail, but nothing turns on the particular features of Yablo's framework. The argument is supposed to apply to *any* view on which conceiving or imagining is supposed to be evidence of possibility—i.e., where telling a story or visualizing a state of affairs is supposed to be evidence that there is a world of which that story is true, or in which that state of affairs obtains.) Very roughly, Yablo's theory goes as follows. If p is conceivable for me, then I can imagine a world that I take to

[6] Van Inwagen doesn't actually give examples, though he clearly has propositions like these in mind. Nor does he explain how we know that claims like these are necessarily false. I tend to agree that they are necessarily false, but I wouldn't use them to object to the thesis that consistency is a guide to (absolute) possibility. As I suggested earlier, I think Pruss's argument, positing abstracta, a posteriori necessities, and semantic indeterminacy are more useful for this purpose.

verify *p*. In the absence of defeaters, I'm justified in believing that things are as they seem. So, if I can imagine a world that I take to verify *p*, then I am justified in believing that there *is* a world that verifies *p*—which, of course, is to be justified in believing that *p* is possible. Therefore, if *p* is conceivable for me, then I am justified in believing that *p* is possible. Van Inwagen latches onto Yablo's first step, arguing that, for any extraordinary proposition, we have no reason to suppose that we've imagined a world that verifies it. He asks us to consider, for example, whether we can imagine a world in which there is transparent iron. His response is that we can't:

> ... unless our imaginings take place at a level of structural detail comparable to that of the imaginings of condensed-matter physicists who are trying to explain, say, the phenomenon of superconductivity. If we simply imagine a Nobel Prize acceptance speech in which the new Nobel laureate [...] displays to a cheering crowd something that looks [...] like a chunk of glass, we shall indeed have imagined a world, but it will not be a world in which there is transparent iron. [...] This sort of effort of imagination will [...] show that a certain proposition has the modal status "possible," but the proposition will be a disjunctive one. Here are some of its disjuncts:
>
> - Transparent iron exists
> - The scientific community has somehow been deceived into thinking that transparent iron exists
> - A crackpot physicist who thinks he has created transparent iron is the butt of a cruel and very elaborate practical joke
> - A group of fun-loving scientists have got together to enact a burlesque of a Nobel Awards Ceremony.
>
> [...] No doubt, by working our imaginations a bit harder, we could imagine a world in which some of the "unwanted" disjuncts failed. We might, for example, add to what we have already imagined a codicil to the effect that all the scientists in the cheering audience are sincere. But this would not rule out the second of the above disjuncts ("mass deception"). To rule that out, our imaginations would have to descend to "a level of structural detail comparable to that of the imaginings of condensed-matter physicists who are trying to explain superconductivity" (1998, 79–80).

So van Inwagen is *not* contesting the following conditional: if you could indeed imagine transparent iron, you would have some evidence that transparent iron is possible. What he contests is the antecedent of this conditional. His suggestion is that we only imagine transparent iron if our imaginations descend to a level of structural detail comparable to that of the imaginings of condensed-matter physicists who are trying to explain superconductivity. And, of course, that never happens: either the task is too complex or it would require knowledge that we lack. So, we aren't justified in believing that there could be transparent iron.

5.2 Problems for the Disjunctive Argument

Let's call the argument in the last block quote, "the disjunct argument." Heimir Geirsson reads it as insisting that, for any extraordinary proposition *p*, you don't know that *p* unless you can imagine "all the relevant detail in the world that makes

[*p*] true" (2005, 285–288).[7] Peter Hawke (2011) offers a similar interpretation; he claims that the disjunct argument is based on a principle like this: you haven't imagined a world that verifies *p* unless you've imagined a world in sufficient detail so as to rule out ~*p*. (Geirrson's "relevant" and Hawke's "rule out" can be given stronger and weaker readings, and presumably ones that make the two interpretations equivalent.) Both worry that the disjunct argument leads to a more pervasive skepticism than van Inwagen intends. For example, Geirsson considers whether the disjunct argument is compatible with justifiably believing that an LP could play without pops and clicks (a modal claim that he regards as being ordinary, and so one that van Inwagen would say that we know).[8] If being able to do so depends on our being able to imagine all the relevant details of a world that verifies this proposition (or, more carefully, its non-modal counterpart), then I know that I'm sunk: *I* certainly can't imagine in complete detail the technology that would allow you to play an LP without background noise—not least because I can't imagine in complete detail the technology that allows you to play an LP *with* background noise. And, Geirsson reasons, if van Inwagen's argument undercuts our beliefs about uninteresting modal matters, then we have cause to reject it.

This sort of objection misses the mark, and it's important to see why. Van Inwagen never says that our uninteresting modal beliefs are justified in the same way as are our beliefs about extraordinary propositions—not least because he denies that we have justified beliefs about extraordinary propositions. On his view, our uninteresting modal beliefs are either properly basic, inferred from beliefs that are properly basic, or justified by inferences from actuality. There is no mystery in inferences from actuality to possibility. However, if asked to provide an account of properly basic beliefs, he declines: his official position is that, while we know that we have (properly basic) uninteresting modal knowledge, we have no idea *how* we have it:

> Although I do not doubt that we have some modal knowledge, I regard much of this knowledge as mysterious. Some modal statements, I have said, we know by reasoning from what I have called "basic" modal knowledge—simple, obvious modal statements whose truth we are somehow in a position to know—together with some facts about how the world is constructed. But how do we get started in this reasoning? How do we know the "simple, obvious" modal statements to be true? What is the ground of "basic" modal knowledge? I do not know how to answer these questions (1998, 73–74).[9]

[7]For a virtually-identical interpretation, see Cohnitz (2002).

[8]I don't know whether Geirsson is right to say that this claim is an uninteresting one. Here is an argument to the effect that it isn't obviously uninteresting. The following is an uninteresting modal claim: an LP could play with *fewer* pops and clicks. And the following is not obviously an uninteresting modal claim: where *n* is greater than zero, there is no *n* such that an LP could not play with *n*-1 pops and clicks. So, that an LP could play without any pops and clicks is not obviously uninteresting. Part of the problem here may be that van Inwagen doesn't—and so Geirsson and Hawke don't—distinguish between different interesting modal claims, and so if it isn't obviously uninteresting, it gets quickly classified as extraordinary. But there are steps in between.

[9]Perhaps this is a way of saying that our everyday modal knowledge is Moorean. Or perhaps this ignorance is an implication of our everyday knowledge being Moorean. I'm not sure. Whatever the

Now, I don't know whether van Inwagen would grant that our uninteresting modal knowledge does not meet the standard to which extraordinary modal claims are held in the disjunct argument. He could maintain—or at least I see no reason why he couldn't maintain—that our acquaintance with a range of ordinary objects in ordinary circumstances explains why our imaginings or conceivings meet the relevant standard. (I don't know how the explanation would go, but it would be an interesting line to pursue.) Alternately, he could dig in his heels, conceding that our uninteresting modal beliefs do not meet the disjunct argument's standard, but insisting that this just shows them to have some *other* foundation—not conceiving or imagining.

Even if van Inwagen can sidestep Geirsson's objection, it does reveal that we are without a clear argument for the high standard that the disjunct argument employs, and therefore without a clear answer to doubts about the claim that our imaginings are *never* sufficiently detailed to provide evidence that a *p*-friendly world exists. What's more, I think that there are deeper reasons to suspect that van Inwagen cannot provide the needed argument.

Let's suppose that you need to be able to imagine all the details relevant to an extraordinary proposition's truth in order to know its modal status. What might motivate this requirement? Recall the transparent iron example. On the assumption that you aren't simply *stipulating* that you're imagining a hunk of transparent iron, on what basis are you *identifying* it as such? Well, if you only identify *actual* samples of iron on the basis of expert testimony, then you are in no position to take your imagining as evidence of possibility.[10] Trusting your imagining would be like trusting the testimony of a hapless prospector: he can't tell the difference between fool's gold and the real thing; so, his claim to have struck gold should be eyed with suspicion. (And in fact, trusting your imagining would be *much*, *much* worse than trusting the testimony of the hapless prospector: he's got a fifty-fifty shot at being right, whereas there are many, many ways for you to have a false positive.) The "all-the-relevant-details" requirement is supposed to ensure that your imagining has

details, you might wonder whether this view actually works in his favor. If we know that we have modal knowledge, then even if we can't explain it, wouldn't it be natural to suppose that our actual practices are trustworthy (absent defeaters) not just for ordinary propositions, but for extraordinary ones too? I am not so sure. If van Inwagen is right that we can't explain our modal knowledge, then surely we're in a poor position to discriminate genuine cases of modal knowledge from spurious ones. And if that's right, then we should tread very carefully once we move beyond those claims about which we are all quite confident. This is just a burden-shifting move: instead of being able to rely on the presumption that a belief is innocent until proven guilty, van Inwagen's interlocutor is now under some pressure to offer a positive story about how she knows that there is a history and set of laws with which a given proposition is compossible; if she can't, then her claim to know the modal status of a given extraordinary proposition becomes suspect. The upshot is that, if you don't know how you know, then you shouldn't trust yourself when you shift from a familiar to an unfamiliar context (i.e., from the ordinary to the extraordinary); and if you don't know how you know, then you should be much more bothered by disagreement that you might be otherwise.

[10]Moreover, it's no good to stipulate that the expert testimony is correct, since your imagining provides no reason to believe any such thing.

better epistemic credentials: the requisite detail makes it more plausible that you have the ability to identify the thing that you're imagining, and hence makes it more plausible that what *seems* possible *is* possible.[11] So far so good. But what constitutes imagining all the relevant details? I understand this as the requirement that you be able to imagine a scenario *s* such that, necessarily, if *s* were the whole world, then extraordinary proposition *p* would be true. In other words, you need you need to be able to imagine a *model* of *p*. This explains why the Nobel Prize acceptance speech scenario is insufficient, since the description of it is compatible with there being no transparent iron. It also explains why the right scenario would probably be extraordinarily detailed (e.g., as detailed as the model constructed by condensed-matter physicists who are trying to explain superconductivity).

However, once we put the condition this straightforwardly, it quickly falls apart. Whatever a possible world *is*, it is somehow *maximal*, which is to say that (assuming bivalence for simplicity) every proposition is either true or false at that world. But there is no similar restriction on the construction of models: they *always* leave unspecified all sorts of details. So, our scenario *s* may well be consistent, but that it is does not entail (or even suggest) that there is some world of which *s* is a part: as van Inwagen argues, consistency is no guide to possibility. It follows that *p*'s being true of *s* is no evidence that *p* is possible, since it is no evidence that there is some world at which *p* is true. So, even if we could meet the standard in the disjunct argument, the gap between consistency and possibility should lead van Inwagen to deny that we've done enough.

What *would* be enough? Well, I presume that if you could imagine *every* detail—not just every relevant one—then you could have extraordinary modal knowledge. In other words, you would need to imagine *a whole world* that verifies *p*. But of course this is impossible: no one can entertain that many propositions (either simultaneously or sequentially), and to demand as much is unreasonable. Granted, if you are already inclined to regard extraordinary modal knowledge as impossible, then van Inwagen may have a way to explain why it has this unfortunate status. But this is not much of a basis for modal skepticism.

These problems make me suspect that we have not been charitable enough in our interpretation of van Inwagen's position. In conversation, though, van Inwagen assured me that he indeed holds a view like the one that Geirsson and Hawke attribute to him. I have no reason to doubt van Inwagen on matters of van Inwagen

[11] Incidentally, Kripke is often cited as the authority on the legitimacy of stipulation when it comes to matters modal: see his (1980, 44–47 ff.). But Kripke is not really interested in the epistemology of modality in the relevant section of *Naming & Necessity*. Rather, he's making a semantic point: we certainly seem to know that *this very man* could have been different in various ways, and the best semantic theory allows us to say as much (which, according to Kripke, you can't say if you adopt counterpart theory). If we *can* say as much, and if we know that Nixon could have lost the election, then we needn't characterize a world purely qualitatively. Instead, we can stipulate that we're talking about the world in which Nixon loses the election. Even if correct, this observation is no comfort to the defender of stipulation as part of a modal epistemology. More on this in Chap. 6.

interpretation. So, I conclude that van Inwagen's version of modal skepticism faces some serious problems, and that if the spirit of the view is correct, then we need to flesh it out in a better way.

5.3 The "Direct Evidence" Version of Modal Skepticism

Here is an alternate route to modal skepticism that aims to preserve the basic structure of the disjunct argument. Although I think that it fails, its failure is instructive.

When considering a scenario that's *supposed* to verify the proposition, "There is transparent iron," the natural worry is that the imagination doesn't maintain a sufficiently tight grip on the substance in question: perhaps the scenario fails to depict a substance at all, or perhaps it depicts some familiar substance that we confuse for transparent iron (e.g., glass), or perhaps there is no fact of the matter about which of many transparent substances it depicts. Geirsson and Hawke are trying to make sense of how van Inwagen's requirements for the imagination to "hang on" to its object under various transformations. And this is, I think, the right goal, even if we have yet to find a way to reach it.

One possibility is to say that you can't successfully imagine p by imagining circumstances that would, if actual, provide just any old evidence of p's truth. Why not? Suppose that I know nothing about iron's nature and some expert tells me that iron is transparent under such-and-such conditions. No doubt her testimony constitutes evidence for me that there are (or could be) samples of transparent iron. However, the value of her testimony is parasitic on a certain kind of non-testimonial evidence—I'll call it "direct"—that comes in two main forms: first, observation by someone with the appropriate discriminatory abilities; second, suitable confirmation of the theory that predicts that there are such samples. If we want to provide an account of the source of our justification about modal matters, then we should focus on direct evidence. So, if your aim is to imagine transparent iron, then you need either (a) to be able to discern genuine instances of transparent iron from imposters or (b) confirm a theory—based on a thought experiment—that predicts the reality of transparent iron. On the face of it, neither option seems terribly promising for ordinary modalizers.

But the situation may be less dire than it looks. Consider what it would take to convince you that x is a sample of transparent iron. If the transparency is obvious, then the thing to do would be to check for iron's characteristic properties. (If you don't know what those properties are, no matter: just look them up. Isn't that what you'd do normally?) Suppose you find that your sample has all the relevant properties. No doubt you would conclude—or at least *could* reasonably conclude—that you were dealing with transparent iron. Of course, given your lack of expertise, perhaps you shouldn't be wildly confident about your conclusion, and wisdom recommends deference to an expert should one disagree with you. Still, if your evidence points to p's truth, and you have no defeaters, then where's the error in believing?

Yablo's epistemology is based on the parallel between perceiving and conceiving; the presumption, therefore, should be that you can use the same epistemic tools in the imagination that are available in perception. In particular, you should be able to believe whatever would best explain your imaginative experience. So if you can imagine a transparent sample having the properties characteristic of iron, and if its being a sample of transparent iron would best explain that experience, then the analogy seems to give you license to conclude that you're dealing with transparent iron. Note, moreover, that the standard for justified belief is sensitive to context in the following weak way: before you've been challenged, it may take very little imaginative evidence to justify the belief that there could be transparent iron, though you'll have to provide more once your belief has been called into question. But why think that you can't? Surely you can imagine a bubbling reaction when you put a few drops of the right chemical on the sample, or imagine seeing the solution in the test tube change colors when you add some filings to it.

You might worry that this doesn't address one of van Inwagen's central concerns, namely, that if you believe that *p* is possible, then you are committed to there being a history and set of laws with which *p*'s truth is compossible. Again, the analogy with perception is instructive. Consider this argument pertaining to beliefs about actuality:

> If you believe that *p*, then you are committed to *p*'s being consistent with our world's history and laws (since we don't live in an inconsistent world). If you recognize this, then your entitlement to believe that *p* stands or falls with your entitlement to believe that *p* is compossible with our world's history and laws. Now consider some proposition *p* that meets the following conditions: (1) it concerns matters remote from the practical business of everyday life, (2) its truth-value cannot be determined by logic and reflection on the meanings of words or by the application of mathematical reasoning, and (3) its truth-value is unknown to us. You should be agnostic about whether *p*'s truth is compossible with our world's history and laws. Hence, you should be agnostic about *p*.

This argument is ridiculous. If the best explanation of your experience is that *p* obtains, then you are *thereby* justified in believing that the actual world's history and laws are compossible with *p*. If we model modal knowledge on perceptual knowledge, then the beliefs produced by the imagination will be innocent until proven guilty, and they will be aided and abetted by our methods of ampliative inference. Hence, if the best explanation of your imaginative experience is that *p*, then you are justified in believing that there is a history and set of laws with which *p*'s truth is compossible.

In this light, we might compare the disjunct argument to the cartoon-version of Hume's critique of miracles: both are based on an implausibly stringent account of when it is reasonable to form a belief based on appearances. Suppose that a miracle is a violation of a law of nature, that you believe *L* to be a law of nature, and that it seems to you that you witnessed a violation of *L*. Hume (or cartoon-Hume) seems to think that you always have better reason to affirm *L* than to believe that the miracle has occurred. Laws, after all, are supported by uniform experience, and we know that the senses sometime deceive us. One of the standard objections to this argument is that, if correct, we would never be in a position to discover anomalies of *any* kind—miraculous or otherwise. Since we surely can, the argument must be flawed. The problem is that this way of weighing evidence is unduly conservative: it puts too much emphasis on preserving current theories, presumably in the interest of not disrupting the web of belief. But conservatism and consistency are not the only epistemic desiderata: it's also important to follow the evidence wherever it leads, whether or not doing so disrupts the web. The same is true in modal epistemology. If it *seems* to you that there are *p*-friendly worlds, and you aren't aware of any reasons to be suspicious of this seeming, then it would just be dogmatism to resist the belief that *p* is possible. Of course, it may not seem to van Inwagen that there are any *p*-friendly worlds, or he may think that he has a defeater for the seeming. But if the defeater is just that neither he nor anyone else can rule out the possibility of error, then his argument is on poor footing indeed. We are all fallibilists now. If van Inwagen is demanding certainty, then while we should admit that his version of modal skepticism is true, we shouldn't care.

The short of it is this: if we take seriously the analogy between modal and perceptual knowledge, then we seem to be left with an epistemology on which we have (or can have) *lots* of knowledge about extraordinary modal matters. So, van Inwagen can't appeal to Yablo to provide a basis for modal skepticism.[12] However, there remains something plausible about van Inwagen's view: it seems right that we're at great risk of overlooking relevant possibilities, which explains the worry about committing ourselves to an entire world with which some particular conceiving is compossible. We need to preserve this insight, but if there's any basis for modal skepticism, it will have to come from a different epistemology.

5.4 An Alternative Proposal

Summarizing his objection, Geirsson says that "[t]he problem with van Inwagen's account is that it ties justification of *p*, where *p* is the relevant possibility statement, too closely to both our present state of knowledge and our ability to imagine the details that make *p* possible" (2005, 288). In my view, this isn't a bug in van Inwagen's account; it's a feature. However, given the analogy between modal and perceptual knowledge, I doubt that van Inwagen's account delivers what it promises. So, if you want to preserve the key insight that drives modal skepticism—namely, the worry about overlooked defeaters—then you ought to reject the parallel between perceiving and imagining or conceiving. TEM does just this.

[12]My criticisms in this section suppose that there is indeed a useful parallel between modal and perceptual knowledge. I'll question this assumption in Chap. 6, arguing that the perceptual analogy should lead us to deny that imagining provides *any* evidence of possibility. But there are two reasons why this is no consolation to van Inwagen. First, he doesn't want to deny the conditional: *if* we imagine that *p*, then we have evidence for *p*'s possibility. He only wants to deny the antecedent. Second, it remains the case that the considerations that motivate his modal skepticism fit better with TEM than with an imagination-based modal epistemology, as I'll suggest below.

According to TEM, our justification for believing an interesting modal claim—and so an extraordinary one—is closely tied to our present state of knowledge: for any interesting modal claim, we are justified in believing it only if we justifiably believe a theory according to which it's true. Moreover, our justification for believing that claim is no better than is our justification for believing the theory behind it.[13] And according to TEM, our justification for believing any modal claim is tied to our ability to imagine the details that make *p* possible—or more accurately, our ability to *understand* the details that make *p* possible, at least on some level. TEM doesn't require that we (sensuously) imagine one of the models specified by the theoretical definition, but it does require that we justifiably believe a theory about *p*, and that theory will (or should) shed some light on why *p* is true in the world(s) in question. After all, our best theories are the best—at least in part—because they explain why propositions about actuality are true; so, it's plausible that they can also explain why relevant propositions about other worlds are true, and hence provide at least some details about why those propositions are possible or necessary.

Here is an example that puts these features in perspective. Let's suppose that you are competing in the something akin to the Pinewood Derby. The inclined portion of the track is one meter long and it is at a 30° angle to the flat portion. Now, if you were to set your car at the top of the track, is it possible for your car to reach a velocity of 20 m/s before it hits the bottom, assuming that someone no one propels it to that velocity? If TEM is correct, then it's natural to begin answering this question by revisiting our high school physics. As you reduce resistance, what is the limit that the velocity approaches? The details of the answer matter. We begin by recalling that the force of gravity, F_{grav}, equals ($m \cdot g$). That force is perpendicular to the ground, not the angled part of the track; so, we have to decompose F_{grav} to find the car's acceleration: one part parallel to the inclined part of the track ($F_{parallel}$); the other part perpendicular to it ($F_{perpendicular}$). If the angle of inclination is 30°, then a little geometry will reveal that the angle between F_{grav} and $F_{perpendicular}$ is also 30°. This means that $F_{parallel}$ is ($F_{grav} \cdot \sin 30$), which is equal to the net force on the car along the track (because we're ignoring friction). By Newton's second law, $F_{parallel}$ is equal to the mass times the acceleration of the car. So, ($F_{grav} \cdot \sin 30) = (m \cdot a)$. Solving for a (and remembering that $F_{grav} = (m \cdot g$)), we get $a = (g \cdot \sin 30)$. Next, we take a formula from Newtonian kinematics ($v_{final}{}^2 = (v_{initial}{}^2 + (2 \cdot a \cdot d)))$ and plug in (1) the distance traveled (one meter), (2) the initial velocity of the car (zero, since you simply place the car at the top of the track) and (3) the standard value for gravity (9.80665 m/s^2). The result is that the maximum final velocity of the car is (roughly) 3.13 m/s. In other words, TEM would have us say that the car *could not* reach a velocity greater than 3.13 m/s. Absent any other considerations, we should

[13] . . . and it's potentially much worse. Suppose that we have great evidence for a theory, but we aren't sure whether this theory says that *p*; or we think that it does, but we suspect that another theory that we justifiably believe says that ~*p*; or... In such cases, our justification for believing *p* would be worse than our justification for believing the theory that says that *p*.

say that it's *metaphysically necessary* that its velocity be less than (or at most equal to) 3.13 m/s before it hits the flat portion of the track—a far cry from 20 m/s.

Our theory gives us some insight into why the car couldn't reach the relevant velocity: once we recognize the relationships between the various factors that influence the final velocity—the force of gravity, the angle of the track, the initial velocity, etc.—we see why some final velocities are possible and other final velocities are not. In a sense, this confirms Geirsson's worry about our being required to imagine the details that make *p* possible. However, those details are what give us some confidence that our answer is correct. They're all that tethers our judgments about modal matters to reality as we understand it, and without them, it isn't clear why we should be at all confident about how things could and couldn't be.

Granted, someone is bound to complain that the car could have a much higher velocity than the one that I've calculated. Why? Because gravity could be stronger, or because the conservation law (which is assumed but not mentioned above) could be replaced by the *multiplication* law, according to which potential energy converts to kinetic energy at double (or triple, or quadruple) the actual rate. But given TEM, the details above reveal exactly what it would take to justifiably believe that the final velocity could be higher than the one that I derived. If you want to say that it is merely *physically* necessary that the car's final velocity be less than (or perhaps equal to) 3.13 m/s, but that it's possible *simpliciter* that the velocity be higher, then you need to justifiably believe a theory (presumably a metaphysical one, but we needn't settle that here) according to which a proposition like one of the following is possible:

- F_{grav} takes some value other than 9.80665 m/s^2; or
- Newton's law postulates a different relationship between force, mass, and acceleration; or
- the principles of kinematics suggest some other relationship between initial and final velocities, acceleration, and the distance travelled.[14]

So if, for example, you justifiably believe a theory according to which Earth's gravity could be stronger than it is, then you are free to use that theory to recalculate the highest possible velocity. The trick, of course, is to provide an argument for such a theory. It won't suffice to describe an alternate configuration of laws and constants that produces the relevant effect—according to TEM, you need an argument to think that configuration is either (a) actual or (b) one that's realized in a model of a theory that we justifiably believe. Neither option seems promising.

Someone will be inclined to insist that all laws are contingent, and so these laws are. There are two problems with this move.

The first is that it will be hard to give an argument for that theory about physical laws. Supposing that TEM is correct, you can't simply rely on intuitions to the effect that various physically-impossible events are indeed metaphysical possibilities.

[14]These are not the only options: you could also look for alternate geometries, or mathematics, or logics, or what have you. But lists have to end somewhere.

What, then, will be the data that's best explained by the contingency of all physical laws? Not actual phenomena, since they will be explained equally well by the hypothesis that the physical and metaphysical possibilities are coextensive. And for the same reason, it's doubtful that any modal phenomena postulated to explain actual phenomena will do the trick either. We may, for example, need to postulate the possibility of unobserved cats to explain (a) the disappearance of a cat behind a piece of furniture and (b) the subsequent appearance of a remarkably similar cat on the other side. However, it isn't clear why this modal phenomena—or more complicated variants—would require postulating physically-impossible possibilities.

The second problem is that even if you're justified in believing that the physical laws are contingent, you aren't therefore justified in believing that any particular physical impossibility is metaphysically possible. For all you know, a car's velocity could exceed 3.13 m/s, but only by some small amount. For reasons unknown, perhaps the metaphysical limit is 4 m/s—a far cry from the 20 m/s mentioned earlier. So until we have an argument not just for contingency, but for *radical* contingency, we have little basis for making particular judgments about possible velocities. Again: what data is best explained by that hypothesis?

So it's a consequence of TEM that if you don't justifiably believe a theory that says that gravity could be stronger (or that something else could be different that would affect the velocity of the car), then you aren't justified in believing that the velocity could be any higher than is suggested by the calculations above. I doubt that any of us justifiably believe a relevant theory, which means that TEM can certainly support modal skepticism.[15]?

5.5 Benefits of the Alternative

I think that a TEM-based version of modal skepticism improves on van Inwagen's in at least three ways. First, it provides a better account of what it is to be an extraordinary proposition. Recall van Inwagen's conditions:

1. it concerns matters "remote from the practical business of everyday life,"
2. its truth-value "cannot be determined by logic and reflection on the meanings of words or by the application of mathematical reasoning," and
3. its truth-value is "unknown to us or [is] known to be false" (1998, 74, 84).

Van Inwagen puts Condition (1) more strongly than he needs to (and perhaps more strongly than he even intends to): modal skepticism needn't be tied to the

[15]TEM *doesn't* imply that you need to know the physics just sketched in order to justifiably believe other modal propositions about our derby: e.g., that car could be moving at a velocity greater than 1 m per *month*, or that the car could be moving fast enough to reach the end of the flat portion of the track. Other, less precise theories (perhaps even folk physics, or at least folk heuristics for actual physics) probably suffice to justify these modal claims. Likewise, TEM doesn't say that you need to know the physics just sketched to *rule out* the possibility of the car's velocity being 20 m/s; again, less precise theories may do the trick.

business of everyday *life*, if this refers to the kinds of activities and interests that we have in common; rather, it can be tied to the business of everyday *lives*, so that specialists can have modal knowledge that is unavailable to the rest of us. It would be nice to able to grant, for example, that physicists and day traders have the modal knowledge appropriate to their respective fields, but not the knowledge appropriate to each other's fields (hobbyists and the bivocational aside). TEM's version of modal skepticism can manage this, since it's plain that we don't all believe the same theories, and so don't all have the same modal knowledge available to us. It allows us to give an account of extraordinary propositions relative to persons and times: p is an extraordinary proposition for a subject s at time t if (a) p isn't true according to any of the theories that s justifiably believes at t and (b) p isn't true according to any of the theories that anyone *else* justifiably believes at t.

When van Inwagen gives examples of extraordinary modal claims—e.g., it's possible that I exist and nothing material exists—it seems to be implicit that these claims are beyond the scope of current human knowledge. We just don't know what we'd need to know to settle whether they're possible. TEM's account of extraordinariness captures this, but it also allows that we can be wrong about whether a claim is extraordinary. I might think that "There could be naturally purple cows" is an extraordinary modal claim. But perhaps there are animal scientists who justifiably believe theories according to which this claim is true. If so, then although "There could be naturally purple cows" seems extraordinary to me, it isn't, as the definition's second condition isn't satisfied.

It's also worth noting that, given this account of extraordinary modal claims, it's trivially true that we aren't justified in believing them. TEM says that if we're justified in believing an interesting modal claim—whether technological, scientific, or extraordinary—then we're justified in believing a theory according to which it's true. But the extraordinary modal claims are just those that aren't true according to any of the theories that we justifiably believe. So, in effect, I've stipulated that we aren't justified in believing them.

From the outset, though, it should have been clear that nothing hangs on the terms "technological," "scientific," or "extraordinary." These were always rough and ready ways to help us appreciate the complexity behind certain claims, and so to call attention to the special expertise that's required for assessing them. That's why I haven't bothered to clarify the difference between "technological" and "scientific" modal claims. Moreover, TEM and the associated account of extraordinariness do not settle, all on their own, whether anyone is justified in believing it to be possible that I exist and nothing material exists, or whether that claim is extraordinary. We can't settle that matter without looking to see whether anyone justifiably believes a relevant theory. Of course, if no one *does* justifiably believe the relevant theory—as van Inwagen probably thinks—then it will be perfectly appropriate to classify "It's possible that I exist and nothing material exists" as an extraordinary modal claim. The TEM-inspired account of extraordinariness is helpful precisely because it brings out the connection between the absence of relevant theory and the identification of a claim as extraordinary. Claims are extraordinary because we don't know enough, just as van Inwagen suggested.

Setting all this aside, let's move on to the second way in which a TEM-based version of modal skepticism improves on van Inwagen's. Given that we don't have good theories about extraordinary modal matters—which is a safe assumption, else they wouldn't be extraordinary—the basic consideration favoring TEM also favors a TEM-based modal skepticism. As I said earlier, the basic worry when assessing a modal claim is that we're overlooking another with which it's incompatible. When we consider ordinary claims, we have a large reference class of similar cases on which to draw. We've seen many tables in many different locations within rooms; nothing seems to preclude their being in one spot rather than other. Having encountered so many tables in so many locations, it becomes implausible that *this specific table* couldn't be elsewhere. The absence of evidence for a conflicting modal claim—some countervailing necessity—is evidence of its absence. However, we don't have relevant cases on which to draw when we consider many technological, scientific, and metaphysical modal claims. Here, our evidence for our theories about the relevant phenomena is the only evidence we have for the absence of conflicting modal claims. That's why TEM is a plausible standard to which to hold extraordinary modal claims. At the time, *absent* theories about the relevant phenomena, we have no reason to suppose that there aren't conflicting modal claims, and we should suspend belief. Again, given that we don't have good theories about extraordinary modal matters, that's why modal skepticism is plausible.

The third advantage is that TEM can handle an objection to modal skepticism that van Inwagen cannot. The backbone of van Inwagen's argument for modal skepticism is the following claim: you justifiably believe that p is possible only if you justifiably believe that there is a p-friendly world—i.e., that there is a history and set of laws with which p's truth is compossible. But how do we know that *every world* has a history and set of laws? In other words, how do we know that it is *necessary* that there be a history and set of laws? This claim isn't an extraordinary proposition *per se*, since its non-modal counterpart is not unknown to us or is known to be false—i.e., we know that our world has a history and set of laws (or, at least, I'm glad to grant as much). However, the claim looks to be an extraordinary proposition, since (a) it concerns matters remote from the business of everyday life and (b) its truth value cannot be determined by logical or mathematical reasoning or from the meanings of words. So, it is not one that we would expect van Inwagen to claim to know. But claim to know it he does, and without it his argument falls apart. What can he say in his own defense?

Nothing comes to mind. However, TEM is not in the same position. According to TEM, we are justified in believing that there must be a history or set of laws only if we justifiably believe a theory that says as much. *Perhaps* we justifiably believe a theory according to which there are worlds without the passage of time, and hence without histories. (We might call these 'solid-state' worlds.) I doubt that this is so, but I don't want to argue about it here. What we do *not* believe is a theory according to which there are worlds without laws. If TEM is correct, then the argument for this is simple. If we are justified in believing that there is a world without laws, then we justifiably believe a theory that has a lawless model. By definition, a theory's models are just those models that satisfy its theoretical definition. And the theoretical definition represents (at least some of) the laws that hold in the models that satisfy it.[16] Hence, if TEM is correct, then you are not justified in believing that there is a world without laws, since there is no model of any theory that corresponds to this alleged possibility. It is, moreover, quite plausible that the real work in van Inwagen's argument is being done by the laws, and not the history. So, it's plausible that TEM can salvage van Inwagen's argument for modal skepticism, whereas his own epistemology leaves it in rough shape.

5.6 Some Cases

Before wrapping up this chapter, I want to look more closely at TEM's epistemic caution by considering two philosophically-interesting cases. By showing how van Inwagen's modal skepticism and a TEM-based version differ, it becomes easier to see why it would be hard to avoid modal skepticism if TEM is correct. But in particular, I want to take seriously the possibility that we might be justified in believing some metaphysical theories, and yet still argue that we won't get all the philosophically-interesting modal claims that you might expect. It isn't that modal skepticism is hard to avoid given skepticism about metaphysics. Rather, it's that even if we justifiably believe some metaphysical theories, they might not be as fine-grained as we'd thought.

5.6.1 The Problem of Evil

First, let's consider van Inwagen's objection to Richard Gale's (1996) version of the evidential problem of evil. Here is Gale's argument. Let *g* be the proposition that God exists. Let *e* be the proposition that details every apparently gratuitous evil of which we're aware, and suppose further that *e* includes all we know about each evil's causal antecedents and consequences, at least insofar as that information is relevant to whether the evil's occurrence secured some greater good or was necessary to prevent some greater evil. Gale's contention is this: you can infer from *e* that, probably, there is at least one gratuitous evil; and, relative to our background knowledge *k*, the probable existence of at least one gratuitous evil lowers *g*'s probability. In short: $P(g/(k \& e)) < P(g/k)$.

Van Inwagen objects as follows. Even if we're justified in believing that there probably *are* some gratuitous evils, we should assign a lower probability to God's existence only if our background knowledge includes the proposition that God could

[16] Or, if you prefer, the theoretical definition represents (at least some of) the invariant features of those worlds, and hence (at least some of) the features that determine what's possible in those worlds.

have prevented them.[17] And, he argues, we have no reason to suppose that God could have done any such thing; we just don't know what probability to assign to the proposition, "God could have prevented there from being any gratuitous evils." I'm not going to discuss van Inwagen's argument for this conclusion, since (a) it is very similar to his argument against our knowing that there could be transparent iron and (b) I've already said what I want to say about that general approach. Instead, I want to show how TEM reaches the same conclusion via a different route.

The crucial question is this: do you justifiably believe a theory that entails that God could have prevented there from being any gratuitous evils? If you don't, or if you do and don't realize that the theory has this implication, then you aren't justified in believing as much, at least by TEM's lights. Well, what theory could do the trick? Presumably, the right theory about God's power would suffice. Given a suitably generous account of the limits of omnipotence, it might follow that God could have prevented there from being any gratuitous evils. But how does this theory work, and whence your justification for believing it?

There are two points to make here. First, as is well known, it won't do to have a theory that implies that God can do absolutely *anything*, since that leads to a contradiction: if God can do absolutely anything, then he can create an unliftable rock; if he can create an unliftable rock, then there is something that he can't do (i.e., lift the unliftable rock), and hence he can't do absolutely anything. So, at bare minimum, the theory needs to be restricted enough to avoid this implication. Second, consider what might count as evidence for a theory of God's power. Presumably, it's going to be the data that drive some version of either the cosmological or teleological argument: e.g., the existence of contingent objects, or the causal powers of contingent objects, or the fine-tuning of the universe for life, or what have you. (Forget about the *success* of any version of the cosmological or teleological argument: justification is relative to your epistemic situation, so someone might justifiably believe a theory about the scope of God's power on the basis of an utterly hopeless argument, assuming that she doesn't see its flaws.) But it's hard to see how any of this data will provide—or even give you reason for an opinion about—the answer to the question at hand.

Suppose you justifiably believe that God is responsible for the existence and arrangement of all contingent objects. What should you infer, on this basis, about the contingent objects that God could have created, or about the arrangements in which he might have placed them? Not much, or so it seems to me. Assuming that the data supports the existence of God, it clearly provides a reason to think that God is very powerful. But we know that the theory shouldn't attribute completely unrestricted power to God. So, any attempt to flesh out "very powerful" has to obey this stricture. But what other guidance does the data provide? Remembering the problem with appeals to "logical possibility," you can't say that God can do whatever's logically possible: that lets in too much. You might say, then, that God can do whatever's metaphysically possible (possible *simpliciter*). This is, no doubt, one of the more

[17] . . . and would have. For simplicity's sake, I'll suppress this extra condition in what follows.

elegant hypotheses available, and might be preferable in part on those grounds. But this theory is *epistemologically* useful only insofar as you have justified beliefs about what's metaphysically possible.[18] On what grounds do you believe it to be possible that there could be a world *sans* gratuitous evils? What theory has this consequence, and on what basis do you believe it?[19] Without answers to these questions, you won't be able to make any inferences about God's ability or inability to create a world *sans* gratuitous evil: at least with respect to this issue, the scope of his power will be inscrutable to you.

N.B., it is no small feat to answer these questions. To choose just one of the many ways to illustrate this point, recall that natural disasters are commonly cited as instances of gratuitous evils—the 1755 Lisbon earthquake being a favorite of several authors. So, a world without gratuitous evils will be a world in which no natural disaster adversely affects human beings. What theory do you justifiably believe that implies that either (a) there is a world containing humans in which there are no hurricanes, earthquakes, floods, or volcanic eruptions or (b) there is a world containing humans *and* hurricanes, earthquakes, floods, or volcanic eruptions, but humans are never affected by them? None comes to mind. It follows that we are in no position to claim that there could be a world *sans* gratuitous evil, and hence in no position to claim that God could have actualized such a world. And this, of course, is just the conclusion that van Inwagen recommends.[20]

A few reflections. You might suppose that the above supports the view our theory about God's power is vague or in some way incomplete. I'm inclined to say, rather, that we don't have a theory of omnipotence at all. One reason for saying this is due to the fact that it's very hard to articulate a theory of omnipotence without explicitly using modal notions, as if being-able-to-do-what's-metaphysically-possible is a property that has some serious explanatory power. In my view, however, this claim is doubtful.[21] Suppose, then, that we have what counts as a sketchy theory of the scope of God's power. The problem is that the above theory of God's power (God-can-do-anything-that's-metaphysically-possible) doesn't specify, all on its own, more or less determinate scenarios as possibilities-that-God-could-have-actualized. Instead, it relies on our having an independent grip on the notion of metaphysical possibility. Again, then, the *epistemological* payoff of the theory about omnipotence is limited;

[18]For an insightful discussion of this point, see Hanrahan (2009).

[19]This problem doesn't threaten the viability of the theory as an explanation of any facts about actuality. Since the contents of the actual world are actual, you can infer that they're possible. You know, therefore, that it follows from the theory that God could create the actual world.

[20]TEM puts some teeth on one of van Inwagen's more memorable comments. He writes, "if you think that it would be possible to design a planet, and a universe to contain it, that was both capable of supporting human life and contained no earthquakes or tornadoes, I can only point out that you have never tried" (1995, 106).

[21]If I'm correct, then this fact probably diminishes, but does not destroy, the explanatory power of the God hypothesis. We needn't understand everything about the molecules of a gas in order to know that those molecules best explain Brownian motion—though greater understanding would probably make deeper (and hence better) explanation possible.

it only tells you that, for any state of affairs *s* that we know to be possible, God could actualize *s*. And this is hardly enough for Gale's argument. So, TEM sides with van Inwagen against Gale's version of the evidential problem of evil. But more importantly, it identifies what we'd need to avoid this skeptical conclusion: namely, a theory that implies—or a few theories that together imply—that there is a world *sans* gratuitous evil. The challenge is show how we should justifiably believe such a theory (or theories) based on the data available to us.

5.6.2 The Ontological Argument

I'll consider just one more case that van Inwagen discusses: the modal ontological argument. He summarizes it this way:

1. It is possible for there to be a perfect being (that is, a being that has all perfections essentially).
2. Necessary existence is a perfection.
3. So, there is a perfect being (1998, 67).

As I indicated at the outset of the chapter, van Inwagen denies that we are in a position to form a justified belief about (1)'s truth or falsity. He contends, first, that there is no good argument for (1), and second there is no reason to suppose that (1) is the sort of claim that we would be justified in believing without argument. In defense of the first point, he rehashes concerns about inferences from logical possibility; in defense of the second, he draws an interesting analogy.

> ...suppose we call a real number *septiquaternary* if '7777' occurs in its decimal expansion; and let us call a real number *perimetric* if it measures the circumference of a circle whose diameter measures 1. Then
> Possibly, something is septiquaternary and perimetric
> (or, alternatively,
> Septiquaternity and perimetricity are compatible)
> is obviously such that no philosopher has a "right to his opinion" about its truth-value (1977, 391).

This analogy made for a better argument in 1977. Now, a few seconds on Google reveal that it is indeed possible to have something that is septiquaternary and perimetric. (Two fun facts: the first such string begins at the 1589th digit in π's decimal expansion, and there are 84 such strings in the first million digits.) Still, it's certainly the case that *before* you perform a Google search, you have no right to an opinion about whether it is indeed possible to have something that is septiquaternary and perimetric. But (1), he argues, is just like "Possibly, something is septiquaternary and perimetric," and since we have no right to an opinion about the one, we have no right to an opinion about the other. Van Inwagen doesn't explain what the similarity consists in, but it's plausible (a) that it isn't up to us whether either proposition is true, (b) that neither proposition matters much for the business of everyday life, and (c) that both propositions concern matters that are fairly complex. Hence, it's plausible that neither proposition is of the sort on which we should venture an opinion without a tolerably good argument.

I think that van Inwagen's case against (1) missed the mark. Suppose that you justifiably believe the conclusions of Aquinas' Third and Fourth Ways (which are supposed to demonstrate the existence of a necessary being and a maximally perfect being, respectively), and you justifiably believe that these conclusions are true of one and the same being. (Again, it doesn't matter whether the reasoning is actually sound; what matters is whether someone could justifiably believe that it is, and so justifiably believe the conclusion. And I suspect that someone could: Aquinas, for example.) Now, if you justifiably believe that there is a maximally perfect necessary being, then no doubt you could be justified in believing that it is *possible* there be a maximally perfect necessary being; even a very weak modal logic will sanction that inference. The problem with the ontological argument is not, therefore, that no one is in a position to take a stand on (1)'s truth. Rather, the problem is that the people who *are* in that position don't need the modal ontological argument, and those who aren't in that position won't be convinced by it. In other words, (1) is the sort of claim that only a theist would believe (and not all theists at that), and so the argument has no force. If an atheist were ever disposed to concede (1), that generosity would surely (and rightly) vanish as soon as she saw that necessary existence is being taken to be a perfection. Since the atheist doesn't think that a maximally perfect necessary is actual, she therefore won't think that one is possible.

TEM can account for this straightforward problem with the modal ontological argument. No atheist should affirm that it is possible for there to be a perfect being, since no atheist should accept a theory according to which that proposition is true. And no theist will affirm that it is possible for there to be a perfect being if he doesn't believe a theory according to which that proposition is true (Richard Swinburne (2004) and Keith Yandell (1988), for example, are theists who appear to reject (1) for just this reason). Moreover, if a theist *does* justifiably believe such a theory, then she already justifiably believes a theory according to which there exists a maximally perfect necessary being; in this light, it's hard to see what the modal ontological argument adds to her justification for theism.

In any case, if van Inwagen offers the wrong diagnosis of the problem with the modal ontological argument, this does nothing to show that his skepticism about (1) is misplaced. The most plausible theory according to which (1) is true is probably one that tries to explain the existence of contingent, imperfect beings—a theory that Aquinas tries to motivate in his Third and Fourth Ways. But elsewhere, van Inwagen argues rather forcefully that the cosmological argument (of which the Third Way is one instance) does not succeed.[22] First, the argument depends on some version of the Principle of Sufficient Reason (PSR), which faces a number of well-known challenges. Perhaps the most formidable is the threat of modal collapse. Let's say that C is the conjunction of all true contingent propositions. C is true if its conjuncts are true, and C is contingent if even one of its conjuncts is contingent. So, C is a

[22] See, e.g., van Inwagen (2009).

contingent truth.[23] By the PSR, there is an explanation of C's truth. If the explanans is a contingent truth, then it is either C or one of C's conjuncts. But no contingent truth explains itself, so either way, the explanans cannot be contingent. On the other hand, if the explanans is a necessary truth, then C is necessary, since the explanans entails the explandum. Therefore, if the PSR is true, then C is necessary, and hence there are no contingent truths.

Second, even if the modal collapse problem can be addressed, the cosmological argument faces another formidable challenge. In the face of the modal collapse problem, the standard move is to weaken the PSR: you deny either (a) that no contingent truth explains itself or (b) that the explanans must entail the explanadum. I see no hope for (a); (b) is the only option. But if you opt for (b), then it becomes much, much harder to show that Explanation X is better than Explanation Y:

X: Contingent beings exist because a necessary being caused them to exist.
Y: Each world is as likely to be actual as the next. So, it's true that *there are some contingent beings or other* because one world had to be actual, ours is that world, and our world contains contingent beings.[24] And it's true that *there are these contingent beings as opposed to some others* because one world had to be actual, ours is that world, and our world contains these contingent beings.

Plainly, Y's viability depends on the claim that each world is as likely to be actual as the next. But there is nothing suspect about this claim. First, we are considering two pictures of how things are: one—X—according to which there is at least one necessary beings who is an agent (or very much like an agent), and another—Y—according to which there are no such necessary beings.[25] If Y is true, then it's safe to say that the only necessary beings are inert objects—like abstracta—and hence not the sorts of objects that can affect the probability of which world is actual.[26] That is, they are not like God, whose character might make it extraordinarily more likely that

---

[23] I recognize that this formulation leads to a version of Russell's paradox, but there is no reason to complicate the story to avoid it since the details would only obscure the main point.

[24] A stronger non-theistic explanation is available for this proposition, though not for the next one. Contingent beings exist in every world but one; so, the probability of there being contingent beings is very high. (Indeed, if n is the number of worlds, then probability of there being contingent beings is $(n\text{-}1)/n$.) So, it's true that *there are some contingent beings or other* because it is highly likely that a contingent-being-world is actual. (Why think that there is only one world without contingent beings? Well, we are considering two pictures of how things are: one—X—according to which there is at least one necessary beings who is an agent (or very much like an agent), and another—Y—according to which there are no such necessary beings. If Y is true, then it's safe to say that the only necessary beings are inert objects—like abstracta—whose (non-Cambridge) properties do not change from world to world. So, all the worlds without contingent beings are exactly alike in every respect. But there is no reason to postulate multiple worlds that are exactly alike in every respect; one will do. Therefore, according to Y, there is only world without contingent beings.

[25] Note that Y does not rule out the possibility of supernatural agents *per se*; it simply rules out the possibility of *necessary* supernatural agents.

[26] ... except by ruling out some worlds as having zero probability of being actualized—namely, the impossible worlds.

some worlds would be actualized over others. Second, it is no good to say that the relative complexity of some worlds makes them less likely than others: a random sequence of heads and tails is more complex than an equinumerous sequence of heads only, but the sequences are equally likely. Third, there is not, of course, a force "outside" the worlds that pressures some toward actualization at the expense of others. So, it's plausible that each world is as likely to be actual as the next. And given this, I see no reason to prefer X over Y. Indeed, you might think that Y is the more plausible story due to its simplicity.[27]

From the perspective of TEM, then, the modal ontological argument is doubly flawed. On the one hand, only those who are justified in believing its conclusion might be justified in believing its first premise. On the other, it isn't clear that there is a good argument for its first premise, at least if the cosmological argument is supposed to do the work. Still, a TEM-based modal skepticism makes it clear what it would take to resurrect the modal ontological argument. You'd need to show that even non-theists justifiably believe a theory according to which there could be a maximally perfect being. The odds are against success, of course, but at least the standard of success is clear and well-motivated.

5.7 Conclusion

Given TEM, a version of modal skepticism isn't inevitable, but it's hard to avoid. In short, this is because it's difficult to argue for the theories that would provide extraordinary modal justification—in van Inwagen's, case-based sense of "extraordinary." However, it's important to note that the difficulty is partially independent of TEM. As I've already indicated, some of the blame lies with TEM's insistence that we can't appeal to untethered intuitions about what's metaphysically possible as data for theorizing about metaphysical modality. However, as we've seen by considering the problem of evil and the ontological argument, the difficulty also stems from familiar challenges associated with arguing for metaphysical theories—namely, underdetermination of theory by evidence and the inherent complexity of metaphysical questions. Those problems remain even if TEM doesn't deserve our allegiance. So, if TEM is plausible, then TEM's caution about extraordinary modal claims might be plausible too.

---

[27]What's more, if Y is in fact a better explanation than X—and the Condorcet winner generally—then we should probably say that (1) is false. On another note entirely, a different theory that might underwrite (1) is the Medieval doctrine of the Transcendentals. On this view, being and goodness are identical; hence, having "maximal being" (which entails, though is not the same as, necessary existence) involves having "maximal goodness" (which I take to be equivalent to being maximal perfect). Insofar as this view requires the existence of God, any evidence for it would be evidence for (1) as well. But I don't know what that evidence could be—or, rather, I don't know what the evidence could be if the truth if theism is not part of our background knowledge.

A non-skeptical version of TEM remains available: you just need an optimistic epistemology for metaphysics, one according to which we're justified in believing various metaphysical theories *without* relying on untethered modal judgments (i.e., modal judgments that aren't themselves justified via theories). But once we set aside untethered modal judgments, I suspect that our epistemology will become much more cautious. Again, this is because I doubt that any actual phenomena—or modal phenomena postulated to explain actual phenomena—is best explained by appealing to physically-impossible metaphysical possibilities. If I'm wrong about that, however, then a non-skeptical version of TEM is in the offing.

Chapter 6
Against Optimism

Abstract Many metaphysicians will be inclined to see TEM as too cautious, and count that as a cost. However, I doubt that TEM's rivals can underwrite more modal justification than TEM offers. To make this case, I examine three representative modal epistemologies—Stephen Yablo's defense of conceivability as a guide to possibility, David Chalmer's modal rationalism, and Timothy Williamson's counterfactual theory.

Perhaps no philosopher is as confident as Richard Gale in the scope of our modal knowledge. To illustrate this, consider his reaction to van Inwagen's modal skepticism:

> I not only disagree with van Inwagen's [skeptical] modal intuitions but fear that he suffers from a serious modal affliction, in spite of his being an excellent philosopher from whom I have learned much. In *On the Nature and Existence of God*, I tried to resolve my modal disagreements with the likes of van Inwagen and Phil Quinn through my modal intuition bowl, but it proved a bust. The networks dropped us because there wasn't enough violence, just a bunch of out-of-shape guys, with the exception of Al Plantinga, who looks like he can jump tall mountains in a single bound, staring at each other and emphatically asserting back and forth, "It is possible that *p*," "No it isn't!"
>
> I have decided to take a more radical, therapeutic approach and have founded the EMDS (Extreme Modal Deficiency Syndrome) Foundation. EMDS is no respecter of rank or philosophical orientation—even tenured analytic philosophers have come down with it. It is tragic to realize that there are people like Peter (my poster person) who cannot modalize as normal people do, or, as we at the Foundation prefer to say, are modally other-abled (1996, 213).[1]

Gale's confidence might be unusual, but the basic view isn't: many would be sympathetic to the view that it's plain that we have some extraordinary modal knowledge (in van Inwagen's sense of that term), and any modal epistemology suggesting otherwise is therefore implausible. Let's call this *optimism* about the scope of our modal knowledge.

Recall the dialectic so far. In Chap. 4, I argued that TEM scores quite well on three virtues: the ability to handle disagreement, simplicity, and conservatism.

[1]Gale's reply is not to van Inwagen (1998), but to two earlier articles. Gale's rhetoric may explain, at least in part, the tone of van Inwagen's (1998) rejoinder.

B. Fischer, *Modal Justification via Theories*, Synthese Library 380,
DOI 10.1007/978-3-319-49127-1_6

However, I also conceded that TEM is a *cautious* modal epistemology, one that doesn't grant us extensive modal justification. In the last chapter, I explored how TEM's cautiousness relates to van Inwagen's modal skepticism, and I argued that TEM can offer a better, more principled version of that view. But plainly, many metaphysicians are optimists, as Gale is, so they won't regard this as a mark in TEM's favor. Instead, they'll be inclined to see TEM as too cautious, and count that as a cost.

However, I doubt that TEM's rivals can underwrite optimism. That is, I doubt that there is another modal epistemology that isn't either (a) at least as cautious as TEM about the scope of our modal justification or (b) far too liberal by anyone's lights. And however confident someone might be about extraordinary modal claims, that confidence shouldn't be weakened by the lack of theoretical support. So, if I can show that other modal epistemologies fall into one of these two categories, that should go a long way toward neutralizing any concern about TEM's cautiousness. (Granted, this means that insofar as TEM's cautiousness is a virtue, it isn't one that distinguishes it from some of its competitors.[2] However, I can live with that here, since my main goal in this chapter is only to show that TEM isn't *worse* than its competitors, at least on this metric.) Of course, I can't go through every account available. So, I'll just examine three representative ones—Stephen Yablo's defense of conceivability as a guide to possibility, David Chalmer's modal rationalism, and Timothy Williamson's counterfactual theory.

6.1 Yablo

Many philosophers seem to think that if we can conceive that *p*, then we are justified in believing that *p* is possible.[3] (Call this "the conceivability thesis.") Steven Yablo has one of the most influential account of this kind. He contends that conceivability is a guide to possibility because, on his view, genuine conceiving is analogous to genuine perceiving: "Just as to perceive that *p* is to be in a state that (i) is veridical only if *p*, and that (ii) moves you to believe that *p*, to find *p* conceivable is to be in a state which (i) is veridical only if possibly *p*, and (ii) moves you to believe that *p* is possible" (1993, 7). The idea is this. Perception presents itself as a guide to its domain. It is reasonable to believe that things are as they seem in the absence of defeaters. Therefore, it is reasonable to believe that things are as one perceives them. But conceiving also presents itself as guide to its domain. So, it's reasonable to believe that things could be as one conceives them.

There are two oft-cited problems for the slogan, "Conceivability is a guide to possibility." The first is that there seem to be cases in which we can conceive that *p*

[2]TEM remains unique in being *designed* to be cautious, as it's motivated by the game analogy.

[3]A close cousin of this view: if we can *sensuously imagine* that *p*—i.e., picture *p*, or auditorily imagine *p*, etc.—we are justified in believing that *p* is possible. I'll discuss this alternative shortly.

and that $\sim p$, though p is necessary if possible and $\sim p$ is necessary if possible: e.g., we can conceive that Goldbach's Conjecture is true and we can conceive that it's false. Second, there seem to be cases in which we can conceive that p despite the fact that we take it to be impossible that p: e.g., we can conceive of Jones being born to different parents, though we might not think that this state of affairs is possible. No doubt there is some sense of "conceive" such that, if the conceivability thesis were to employ it, these would be genuine problems. I can conceive that Goldbach's conjecture is true in the sense that I can conceive of reading the headline, "Mathematicians say: Goldbach was right!"—and I can conceive of going to the conference and hearing the relevant authorities agree that the proof is sound. This is one of the senses of "conceive" that worries van Inwagen, and of course it isn't of much epistemological value: in just the same sense, I can conceive that Goldbach's conjecture is false. Indeed, Yablo identifies five epistemologically-worthless senses of conceiving, each of which would make the conceivability thesis subject to many counterexamples:

1. p is *conceivable*$_{\text{b}}$ iff it is (not un)believable that p.
2. p is *conceivable*$_{\text{bp}}$ iff it is (not un)believable that *possibly*, p.
3. p is *conceivable*$_{\text{ijb}}$ iff one can imagine justifiably believing that p.
4. p is *conceivable*$_{\text{itb}}$ iff one can imagine believing p truly.
5. p is *conceivable*$_{\text{ep}}$ iff one can imagine believing something true with one's actual p-thought (1993, 26).

Yablo contends, though, that none of these varieties presents itself as a guide to possibility; consequently, each fails to be sufficiently like perception. However, he thinks there is another kind of conceiving that is appropriately correlated with judgments of possibility. He explains this variety of conceivability—and its foil, inconceivability—as follows:

> (CON) [p is conceivable for me if] I can imagine a world that I take to verify p.
>
> (INC) [p is inconceivable for me if] I cannot imagine any world that I don't take to falsify p (i.e., for every world that I can imagine, I take that world not to verify p) (1993, 29).

(CON) and (INC) are supposed to help us understand why conceiving and failing to conceive should be correlated with the appearance of possibility or impossibility. If the correlation holds, then the analogy with perception can be maintained. Let's focus on (CON). According to (CON), if p is conceivable for me, then I can imagine a world that I take to verify p. Yablo points out that imagining can be either objectual or propositional. Propositional imagining is imagining *that*; it is *de dicto* imagining, in the sense that one imagines that some proposition is true. Objectual imagining, by contrast, is imagining *de re*; it is imagining some *thing*. Yablo contends that objectual imagining is typically correlated with the impression that the object could exist. And in both (CON) and (INC), he intends "imagine" be taken in the objectual sense. (Yablo denies that the difference between objectual and propositional imagining is to be cashed out in terms of the use of mental images: "Some philosophers use 'imagine' so that imagining a thing is *imaging* it, that is, conjuring up an appropriate sensory presentation. I do not require a sensory-like image for imagining, and certainly not a distinct such image for distinct imaginings"

(1993, 27 n.55; emphasis in original).)[4] The upshot is this. If *p* is conceivable for me, then it seems to me that I can imagine a world that verifies *p*. If there exists a world that verifies *p*—i.e., if there is a world of which *p* is true—then *p* is possible. We can maintain, therefore, that objectual imagining—and hence conceiving—is correlated with the appearance of possibility; if you objectually imagine that *p*, then it seems to you that the truthmaker for "*p* is possible" exists. So, if it seems to me I can imagine a world that I take to verify *p*, then I am justified in believing that there is such a world, and hence that *p* is possible. Therefore, if I can conceive that *p*, then I am justified in believing that *p* is possible (in the absence of defeaters).

There are various objections that have been leveled against Yablo's view. Some have argued that it relies on antecedent modal knowledge, such as some sort of Lewis-style principle of recombination—see O'Connor (2008, 32–36). Others have claimed that it can't handle negative existentials or temporally-indexed facts—e.g., Sosa (2000, 1–5); Geirsson (2005, 292–293). Still others deny that we can imagine a world in enough detail for the appearance of possibility to warrant belief—Oreste Fiocco (2007, 371–373); van Inwagen (1998, 76–81). Each of these objections has some merit, but none, in my view, is devastating to Yablo's project. If you grant the initial parallel between conceiving and perceiving, then I suspect that can be addressed. Nevertheless, I deny that Yablo's view is viable. Yablo's view relies on something like the principle of credulity or phenomenal conservatism: i.e., the view that if it seems to you that *p*, then you are justified in believing *p*, at least in the absence of defeaters. I don't contest the claim that this principle applies in the case of perception, so the burden on Yablo is just to show that conceiving is sufficiently like perceiving to justify extending the principle. However, he also needs to tell a story about how we identify the things that are the objects of our imaginings, and I cannot see how he will do this without seriously damaging the analogy between imagining and perceiving. And when that analogy is broken, the account flounders.

6.1.1 Imagining Without Images

Suppose—as Yablo clearly does—that imagining is not an essentially sensuous mental process (i.e., that you can imagine without imaging). Now recall those famous debates over transworld identity—the claim that individuals can exist in more than one possible world. Those who weren't sympathetic to transworld identity often raised the worry that if Nixon, say, were to exist in another possible world, we'd have no way of *knowing* that it was Nixon. Since we can only identify

[4]Yablo rejects imagining-as-imaging because he thinks that we can't image all the propositions that strike us as being possible; he seems to be convinced by Descartes' argument to this effect based on the difficulty of picturing a chilliagon. At any rate, it follows from this that although you can use images to objectually imagine a world, you needn't; hence, images play no essential role in the justification of your modal beliefs.

him by his properties, and, *ex hypothesi*, there is no one in the world in question with precisely the same set of properties,[5] we lack the means to single him out. Kripke denied that this is a real problem. Coining a now-famous phrase, he maintained that we do not look at worlds through a telescope. Rather, we *stipulate* that the man about whom we're speaking is Nixon, the reference of that name already being fixed in the actual world, and the identity between Nixion-in-$w_@$ and Nixon-in-w_1 doing the rest of the work.

But if we're in a position to *stipulate* that the man in the situation is Nixon, then surely that's because we already *know* that he is. And, if we don't know that he's Nixon, then it would be folly to claim as much without argument. After all, stipulation seems to be on an epistemic par with supposition, and supposition is clearly no guide to possibility. We often suppose the truth of necessarily false propositions, as we do when considering *reductio* arguments.[6] And if we have no trouble supposing the truth of a necessary false proposition, then it's implausible that supposition carries with it the appearance of possibility; it seems much more likely that supposition is neutral with respect to the appearance of possibility, and is merely a way to investigate the implications of a given proposition. (It's also no good to say that someone is only allowed to run with a supposition until it encounters a defeater, unless Yablo provides a richer theory of encountering a defeater than, say, discovering a contradiction. After all, mere consistency isn't a good guide to possibility—or so I argued in Chap. 3.)

So Yablo shouldn't say that we stipulate the contents of our imaginings. What can he say instead? To answer this question, let's consider our attempt to imagine a world that verifies a proposition about the Queen Mother, and let's grant that we doesn't get to say, from the outset, that the world we're imagining contains the Queen Mother. Remember too that there needn't be any images involved in our imagining, so the identification isn't (necessarily) based on the scene's being Queen-Mother-ish. To get what Yablo wants, we need an account that doesn't rely on the *sensuous* imagination. So either the account relies on background knowledge or it doesn't. If it does, then that knowledge is doing the epistemic work—we now need to ask how we got it. But more importantly, this undermines Yablo's project. The point was to give an account of conceiving on which it's a non-derivative source of justification, not a derivative one. So this route isn't open to him.

Let's suppose, then, that images and background knowledge are out. At this juncture, our "imagining" sounds a lot like intuiting: we consider a proposition about the Queen Mother and it simply strikes us as being true.[7] But if this is the

[5]This is true in two senses. First, when considering counterfactuals about Nixon, it's true that no one in the world in question has the same properties as does the actual Nixon (for the simple reason that we are considering contrary-to-fact scenarios *about* Nixon). Second, if you countenance relational properties and reject duplicate worlds, then you'll take it to be a general truth that no one in any world has precisely the same properties as anyone in any other world.

[6]To my knowledge, this observation first appears in Thomas Reid's *Essays on the Intellectual Powers of Man*, Essay IV, Chapter 1, §IV.1.

[7]...or possibly true, depending on whether the proposition is already modalized.

view, then it won't do for Yablo's purposes either. What we want is an explanation of conceiving in terms of objectual imagining—i.e., in terms of a *de re* mental state. What we seem to have is an explanation of conceiving in terms of a *de dicto* mental state. In other words, what Yablo promised us was an account on which the intentional objects of conceiving are (concrete) objects and properties in certain configurations; what we have is an account on which the intentional objects of conceiving are propositions. The former (arguably) preserves the parallel with perception, but the latter does not.

Perhaps that's not what's intended. Perhaps the idea is that intuiting is indeed a *de re* mental state. I'm not sure how the details of such a view would go. But however they go, I suspect that it will be difficult to preserve the parallel with perceiving. Let me try to explain why.

The view under consideration is one on which (a) we somehow attend to a certain configuration of objects and properties and (b) this attending justifies belief that the configuration is possible. The move from (a) to (b) is supposed to be warranted because attending to the configuration makes it seem to us that the configuration is possible—not just that it isn't *im*possible. What explains the appearance of possibility? What, exactly, do these appearances track? To what feature of our (non-sensuous) imaginings are we reacting? To my mind, if Yablo wants to explain the appearance in a way that preserves the epistemic credentials of imagining, then the explanation should appeal to our understanding of—i.e., our available information about—the objects and properties in question. I am not suggesting that the explanation be in terms of antecedent modal knowledge. Clearly, it's unfair to Yablo to demand that he explain the appearance of possibility in terms of, say, our knowledge of the essential properties of objects. However, it does seem fair to ask for some explanation or other. He might appeal to our recognition of some properties as characteristic of the object, or as necessary for all explanatory purposes, or kind-defining, or what have you.

Still, when it comes to the issues raised by interesting propositions, I doubt that we have the information that will figure in this explanation. To see why, let's think about the *source* of our knowledge of objects. Aristotle, for example, has a theory of perception on which the form of an object literally impresses itself on the perceiver:

> That which can perceive is, as we have said, potentially such as the object of perception already is actually. It is not like the object, then, when it is being affected by it, but once it has been affected it becomes like it and is such as it is (*De anima*, Book II, Chapter 5, 418a3-6).

On his view, the eye receives (is "enformed" by) the form of the distal object—i.e., the eye actually instantiates the perceived properties.[8] And since the form of an object is what determines its modal profile, there is no mystery about how we

[8]There is debate about this point; for details, see Christopher Shields excellent *Stanford Encyclopedia of Philosophy* article, "Aristotle's Psychology." However, the exegetical questions can be set aside here, since the morale that I want to draw from Aristotle's view doesn't require deciding between the various interpretations of his position.

can discern the modal properties of the objects that we have encountered. When we perceive, we take in more information than we consciously process (since we attend to only some aspects of the impression due to the form), and some of that information is already modal (or has clearly modal implications). However, I take it that no modern theory of perception works as Aristotle's does. Though it's true that we take in more information than we consciously process, it's doubtful that we're taking in the forms of the objects we perceive—we aren't in contact with their essences. And absent some reason to think that this information is particularly useful for assessing the essences of objects, we should doubt whether we have the information needed to explain the appearance of possibility. If we don't, then surely the view is in trouble, since it makes a mystery of appearances of possibility, and hence of the knowledge that these appearances (allegedly) underwrite.

Someone might try to dodge this problem by taking *de re* intuition as epistemically primitive: a theory can't be faulted for failing to explain what it takes as fundamental. But I think that this would be a recipe for disaster, and the reason for this is simple enough. If Yablo maintains that *de re* intuition is epistemically primitive—and thereby avoids having to explain what our appearances track—then we will be left without a way to choose between competing modal claims. Suppose that we disagree with Smith about whether there could be transparent iron. On what basis should trust our judgment over his? We could say, "Because we had *our* intuition, not Smith's." If so, then we will want to say either (a) why Smith is not an epistemic peer or (b) why, even though Smith *is* an epistemic peer, we should discount his opinion on the matter. Either way, we can then properly maintain that his disagreement does not provide us with a defeater for our claim. But (a) is not an option, since (presumably) we *share* the faculty of intuition (or perhaps less contentiously, the ability to intuit), and because it's an epistemic primitive, we have no basis to say that his works less well than ours. Nor is (b) an option, since *de re* intuition is not plausibly construed as an analyzable process, some part of which can break down. After all, Smith didn't *reason* to his claim; there is no calculation error to which we can appeal. Smith—like us—was simply left with an impression upon considering a configuration of objects and properties. And, if we have no basis for preferring our intuition over Smith's, then it's hubris to trust our intuition over Smith's, a groundless bias if ever there were one. This is the disagreement problem.

We could solve the disagreement problem by saying that we prefer our intuition because it coheres better with our beliefs, or has certain explanatory virtues, or whatever. But if the justificatory work is being done by these sorts of considerations, then it's unclear why we should take *de re* intuition as an epistemic primitive, as it now seems like an unnecessary and dramatic solution to the problem of modal knowledge. So it seems to me that, if we take *de re* intuition as an epistemic primitive, then we are without a solution to the disagreement problem—or, more carefully, we are without a solution that does not bring into question the wisdom of thinking that *de re* intuition is indeed an epistemic primitive. Moreover, there is plenty of deep and persistent disagreement over extraordinary modal claims. So, whatever the merits of a modal epistemology based on *de re* intuition, it is implausible that it is a theory that can give us knowledge of extraordinary propositions. And that, of course, is enough for my purposes.

Let's take stock. At least where interesting modal claims are concerned, it does not appear as though conceiving can do the work that Yablo needs it to do. The problems stem from the difficulty of knowing what you've conceived. Understood one way, conceiving relies on stipulation; understood another, conceiving reduces to some form of intuition, and whatever the form, it's far from clear that it can underwrite interesting modal knowledge. So, optimists cannot plausibly rely on Yablo's modal epistemology.

6.1.2 Imagining with Images

However, we can give up one of the assumptions that leads to these difficulties—namely, that imagining is *not* an essentially sensuous mental process. What if we suppose instead that imagining *is* essentially sensuous? This won't help with every interesting modal claim—e.g., I'm not sure what image I'm supposed to consider when assessing the claim that God could create an undetectable particle—but perhaps this strategy will help us with other interesting modal propositions—e.g., there could be a three-inch thick sheet of transparent iron. That would be success enough.

The view that imagining is a guide to possibility, where imagining is construed as an essentially sensuous process, has been explored by a number of philosophers: among them, Peacocke (1985), Hart (1988), Shoemaker (1993), Hanrahan (2007), and Kung (2010). Those who endorse this view try to salvage the parallel with perception by claiming that it is your ability to *sensuously* imagine that *p* that justifies your belief that *p* is possible, not your ability to conceive that *p*. With exception of Hart (1988), though, none of these philosophers think that their imagination-based views sanction van Inwagen-style extraordinary modal claims. So let's focus on his position.

Famously, Hume said that "nothing we imagine is absolutely impossible" (Hume 1739/1978, 32; *Treatise*, I.ii.2). Hart agrees. He reaches this conclusion via an analogy with perception. Most philosophers are willing to grant that, if you perceive that *p*, then you have reason to believe that *p*. Hart's modal epistemology extends this principle. On his view, perception is to the knowledge of actuality as the imagination is to the knowledge of possibility. So, if you can imagine that *p*, then you have reason to believe that *p* is possible.

It matters to Hart that the imagination is a sensuous process: it involves mental imagery, broadly construed (to include tactile, gustatory, olfactory, and auditory "imaging"). This is crucial for a variety of reasons. First, Hart thinks that "the natural history" of our beliefs about modal matters includes imagery. In other words, he thinks that the phenomenology of the process by which we acquire beliefs about modal matters includes sensuous representation. Second, the imagination helps us to explain judgments that *p* is *im*possible: he claims that it's our consistent and persistent failure to *picture* an object that is simultaneously red all over and green all over that convinces us that there cannot be such an object. (Of course, we shouldn't judge that p is impossible if we're just too simple to understand p. So in addition to consistently and persistently failing to imagine that p, we ought to try to explain our failure in terms of the nature(s) of the object(s) in question. If we can, then Hart thinks this confirms our initial judgment; if we can't, then our failure says more about us than it does about the world.)

Hart puts his modal epistemology to work in an argument for dualism: if you can imagine that p, then it is possible that p; you can imagine being disembodied; therefore, it's possible that you be disembodied. And, of course, if it's possible that you be disembodied, and if identities are necessary, then you are not identical to your body—or to any part of your body, since one can always rerun the argument for that part.

As I've said, supposing is no guide to possibility: we can suppose the truth of necessarily false propositions for the purposes of running *reductio* arguments. However, you might think that imagining also involves supposition. In Hart's argument for dualism, for example, you suppose that you're still imagining yourself after you imagine away your body. If this supposition can't be justified, then imagining may also be guilty of presupposing what's to be established.

It's worth realizing that the prospect of falling back on some pure imagining, untainted by supposition, is not a good one. Consider Christopher Peacocke's (1985) distinction between imagining and "S-imagining." He says:

> "S" is for "suppose": although S-imagining is not literally supposing, it shares with supposition the property that what is S-imagined *is not determined by the subject's images*, his imagined experiences (25).

The perk of pure imagining, or imagining *simpliciter*, is supposed to be that the content of the image *is* determined entirely by the image itself. But as soon as imagining outstrips the image—that is, as soon as the imagining becomes S-imagining, which it does in interesting cases—the problem returns.

What's more, it's doubtful that there is such a thing as pure imagining. Call to mind an image of Bill Clinton working out Hilary's campaign budget.[9] He's bent over his desk, propping up his forehead with one hand and scribbling numbers on a sheet of paper with the other. One can imagine the scene in a bit more detail (perhaps as if from over his shoulder), taking notice of the sources of the amounts that he's adding and the costs that he has to subtract from the total. Is this a case of imagining *simpliciter* or S-imagining? One way to find out is to try to consider another imaginative project that could employ the same image. Even if one brackets skeptical scenarios (imagining that an evil demon causes one to have experiences as of Bill working out Hilary's campaign budget), any one of which suffices to show that this is a case of S-imagining, it's easy to think of other, non-skeptical examples. Imagine that the budget is already worked out, but

[9]What follows is inspired by Kind (2001).

that Bill is under great stress for personal reasons, and so he writes out copies of budget calculations from memory to calm himself. And, since this imaginative experience will be phenomenologically indistinguishable from the one in which Bill is actually working out Hilary's campaign budget, the question is settled in favor of the S-imagining.[10] It's also easy enough to extrapolate from this case to others, and this provides strong evidence that images and imaginative experiences underdetermine the content of the imagining. There are, therefore, very few cases (if any) of imagining *simpliciter*, and this because the standard for imagining simpliciter (having the image/imaginative experience fully determining the content of the imagining) is incredibly high. Consequently, if the imagination is to be a guide to possibility, then it will have to do so in a way that comes to grips with the stipulative aspect of the imagination. How can this be done?

Hanrahan (2007) offers a natural solution. Instead of stipulating that the image-in-mind is of a tiger, or doing something on par with stipulating, you imagine a suitably rich scene and try to determine the best explanation of your imagined experience. Could there be a tiger with green stripes? Imagine a scene and ask whether, if confronted with it, the best explanation would be that there is a tiger with green stripes. If that *is* the best explanation, then you're justified in believing that you are, in fact, imagining a tiger with green stripes, and so that there could be such a beast.

You might worry that this is no improvement on Hart's account: we inevitably make *some* assumptions about what we are imagining (or we do something that is no better, epistemically speaking), and hence the explanatory inference will always be corrupted. This is probably too quick. No doubt stipulation is involved in generating mental images, but that's compatible with being able to step back and regard those images as data. So, the recipe for calling up the image may look like it begs the question: e.g., "Imagine a transparent bar of iron." But this isn't problematic if it's used just as a recipe for producing the appropriate mental image. Then, you can consider your experience not *as of a bar of transparent iron*, but just *as of this transparent and very heavy substance*. The explanatory inference enters at this point to evaluate the imagined scene, thereby avoiding one of the difficulties that Hart's account faces.

Still, even if modal empiricism improves on Hart's view, it doesn't improve enough: it only makes you slightly more passive with respect to the way that things seem to you, and not nearly as passive as you are in perception. According to Hanrahan, you can acquire justification by imagining because you're justified in believing that things could seem as they seem in the imagination—i.e., you could

[10]This is not to say, of course, that there aren't ways *to make it more likely* that an imaginative experience represents one state of affairs rather than another. One might imagine oneself *asking* Bill what he's doing, whereupon he'd say that he is, in fact, working on Hilary's campaign budget (and not calming his unsteady nerves). Still, Bill might be self-conscious about his unsteady nerves, and so he both lies about what he's doing and destroys his scrawled-copies of the budget upon feeling some relief, thereby eliminating the most obvious indications of the nature of the situation. I take it that imagining increasingly sophisticated investigations can always be met by equally ingenious ways of hiding the evidence.

encounter a scenario that led you to have the experience that you're simulating in the imagination. True, an explanatory inference is crucial to determining how best to interpret a given seeming, and when making that inference, you treat the imagined scene as given. This gives the impression of passivity, but it's still the case that you're fully responsible for how the scene appears in the imagination.

An expert on vision once told me that you can produce the appearance of bursts of light by putting pressure on certain parts of the eye. This, as it turns out, is true. But if you try this on yourself, you certainly shouldn't believe, on the basis of your experience, that there is a carefully-hidden someone pestering you with a flashlight—even if that would normally be the best explanation of bursts of light in the relevant region of your visual field. Why not? The obvious reason is *that you fabricated the experience*, and because you fabricated it, you shouldn't suppose that its significance can be read off its apparent content. In other words, you shouldn't suppose that the way things *seem* is a good guide to way things *are*. You should take instead a more detached view, considering the best explanation of your *having* the experience, and not the best explanation of the content, since it doesn't enjoy the *prima facie* trustworthiness that attaches to normal experience. Moreover, because there is no reason to think that *p*'s possibility is likely to be the best explanation of your ability to imagine that *p*, there is no reason to grant the presumption that things could seem as they seem in the imagination.

None of this is to suggest that we can't acquire modal justification through the imagination. Surely we can. The point is just that, as Hanrahan construes it, the imagination is parasitic on independent information about those experiences we could and couldn't have. Perhaps we have good sources of that information for some modal claims. However, given that we have no story those sources, there is no reason to trust the imagination when it comes to interesting modal claims—as Hanrahan would be the first to concede. So, optimists get no help here either.

6.1.3 The Modal-Knowno Problem

I've been arguing against conceiving and imagining as independent sources of justification for interesting modal claims. Whatever we make of those arguments, there is a more general problem that such accounts face: the modal-knowno problem.

To appreciate the problem, consider a criticism that Peter van Inwagen makes against the modal ontological argument. The crucial premise of that argument is that there could be a perfect being, where a being is perfect only if it exists necessarily. Against this premise, van Inwagen writes:

> ... there can be no presumption in favor of possibility ... To see that this is a fact, consider the concept of a "knowno": the concept of a being who knows that there is no perfect being. There would seem to be no reason, on the face of it, to suppose that there being a knowno is an intrinsically impossible state of affairs, like there being a liquid wine bottle. But consider. If a knowno is not intrinsically impossible, there is a knowno in some possible world. But then there is a possible world in which there is no perfect being, since, if someone

> knows something, then what that person knows is true... It follows that if a knowno is possible, a perfect being is impossible—and that if a perfect being is possible, a knowno is impossible... We have, therefore, a pair of concepts—the concept of a perfect being and the concept of a knowno—such that either one of them is possible only if the other is impossible. And we have no argument for the impossibility of either concept. If we adopted the general rule "A concept is to be assumed to be possible in the absence of an argument for its impossibility", we should have to assume both these concepts to be possible, and we know that it is false that they are both possible (2008, 207–208).

In this passage, van Inwagen's point is that we can't simply *assume* that propositions are possible, since for any proposition you like, you can identify and assume one with which it's incompatible. Fair enough.

That point, however, doesn't cut against conceiving or imagining as a guide to possibility. However, let's now consider whether you can *conceive* of a knowno (ignoring imagining for simplicity's sake, though the same points apply *mutatis mutandis*). Let's suppose you can. If so, then you can also conceive of a *modal*-knowno—i.e., a being who knows, of some proposition *p*, that *p* is impossible. Now suppose it's conceivable that *p*—the very *p* that the modal-knowno knows to be impossible. If *p* is possible, then there can't be such a being; likewise, if there could be such a being, then *p* is impossible. Plainly, you can't be justified in believing that both are possible, and that means that you can't be justified in believing that either is possible. Moreover, this is true for any value of *p*. So, we now have a general argument against conceiving as a guide to possibility: the conceivability of modal-knownos provides an undercutting defeater for *every* inference from conceivability to possibility, which means that conceivability is an unreliable guide to possibility. It isn't just that the presumption of possibility goes by the board, across the board, as in van Inwagen's original argument. Rather, it's that we have to change our default stance concerning the reliability of this mental operation. Instead of being allowed to assume that it's innocent until proven guilty, we ought to assume the converse. This is *the modal-knowno problem.*[11]

Now: either conceiving does or doesn't face the modal-knowno problem. If it does, then we need to offer positive arguments for possibility claims; that they're conceivable isn't sufficient. So suppose conceiving doesn't face the modal-knowno problem. Presumably, this is because (a) there's a special problem with the inference from the conceivability of the modal-knowno to the possibility thereof or (b) we aren't justified in believing that we've conceived of a modal-knowno. I have no idea what the problem might be with the inference, so I ignore that possibility. However, it's easy to see why someone might press the latter point.

[11]To be clear: the concept of modal-knowno is not of a being who knows, for all *p*, that *p* is impossible. Rather, the concept is of a being who knows, of some particular claim, *p*, that *p* is impossible. This explains why it doesn't matter whether, for some values of *p*, the concept of a modal-knowno is self-referentially incoherent. (Consider, for example, the claim that *there is a modal-knowno*.) All this shows is that there couldn't be a being who knew *that particular claim* to be false—not that there couldn't be modal-knownos generally. Indeed, if you think you know that there *couldn't* be such a modal-knowno, then you think you know that a particular claim is impossible; hence, *you* are a modal-knowno.

After all, someone might insist that you can't be confident that you've conceived of a modal-knowno unless you're *already* confident that whatever that being *believes* to be impossible *is* impossible. In other words, despite the apparent conceivability of some particular modal-knowno, you should doubt that your conceiving has been successful unless you are reasonably confident that the conditions obtain that *make that modal-knowno possible.* If you aren't so confident, then there remains a relevant possibility that you've failed to eliminate—namely, that the modal-knowno merely *believes* that the claim is false, and is mistaken about this.

But if this is the standard for conceiving of modal-knownos, it should be the standard elsewhere too. There is nothing special about this case. And, of course, when it comes to interesting modal claims, we can't meet that standard: we don't know enough about the conditions that would need to obtain to make those claims true, and so don't know whether the claims are in fact possible. Conceivability (or imagining, or whatever) might well provide evidence for some ordinary modal claims, but as the modal-knowno problem shows, it doesn't help with the interesting ones.

Let's take stock. Either conceiving does or doesn't face the modal-knowno problem. If it does, then we have reason to think that conceiving is an unreliable guide to possibility, and we need to offer other positive arguments for possibility claims. No one who sympathizes with conceiving as a guide to possibility is going to take this horn of the dilemma. So suppose that conceiving doesn't face the modal-knowno problem. If not, then this is probably because we aren't justified in thinking that we've conceived of a modal-knowno. And if that's so, it's probably because we can't rule out the possibility that we are conceiving of a being who believes falsely that *p* is impossible. To eliminate that possibility, we need to know about the conditions that would need to obtain to make the modal-knowno possible. But we don't have that knowledge concerning interesting modal claims. When it comes to whether there could be swamppersons, or transparent three-inch-thick sheets of iron, or whether minds could exist without bodies, we have no reason to think that the relevant set of properties is compossible.

So insofar as we take *all* our modal knowledge to come by the independent activities of conceiving, imagining, and/or intuiting, we should be modal skeptics. Alternately, we can give up the idea that these are independent sources of justification, in which case we don't know which beliefs they can justify until we hear more about the background mechanisms. TEM, of course, is one such mechanism, and it can lead us to modal skepticism as well. In the next two sections, we'll consider two others. By my lights, they lead to the same place.

6.2 Chalmers

One way to underwrite conceivability is to have it rely on semantic knowledge. This is David Chalmers's strategy. On his version of modal rationalism, ideal primary positive conceivability entails primary possibility. Let's unpack this statement.

First, we can distinguish between prima facie and ideal conceivability. It might be prima facie conceivable that there is a barber who shaves everyone who doesn't shave himself, but it isn't ideally conceivable—on further reflection, we see that such a barber is impossible. This isn't to suggest that ideal conceivability is *merely* what results from further reflection—it isn't secunda facie conceivability. Rather, it's relevantly perfect reflection. Second, we can distinguish between positive and negative conceivability. A proposition is negatively conceivable for a subject if that subject can't rule out *p* a priori. Positive conceivability, by contrast, involves conceiving of a scenario of which *p* is true. I can't rule out the falsity of Goldbach's conjecture a priori, and so it's negatively conceivable for me. However, I can't positively conceive of Goldbach's conjecture being false—i.e., I can't conceive a scenario where it's clear that the conjecture is false, such as one involving the details of a proof showing that it's false).

Finally, we can distinguish between primary and secondary conceivability, which is closely related to primary and secondary possibility. In short, the idea that there are two ways that we can read a claim like "It's conceivable that water is XYZ." On one reading, the claim is that *it could have turned out* that water is XYZ. That is, we could have lived in a world in which, given one dimension of the meaning of "water"—i.e., its primary intension—it would have referred to XYZ. Water's being XYZ is primarily conceivable in this sense. However, it isn't secondarily conceivable. On this reading, the claim that "It's conceivable that water is XYZ" says that, given how things actually are—i.e., what "water" refers to in the actual world, which affects its secondary intension—it's conceivable that water be XYZ. And that, it seems, is mistaken for familiar Kripkean reasons: once we fix the reference of "water," it refers to the same chemical kind in all worlds, and so water can only be H_2O. In other words, it's conceivable that water is XYZ if we consider the XYZ-world as actual; however, it isn't conceivable that water is XYZ if we consider that claim counterfactually, where it's given that "water" refers to H_2O in in the actual world. Primary and secondary possibility are just the senses of possibility that correspond to the relevant senses of conceivability—i.e., it's primarily possible that water be XYZ, but not secondarily possible.

We can now appreciate the claim that ideal primary positive conceivability entails primary possibility. If Chalmers is correct, then conceivability theorists can navigate the problem of a posteriori necessities. On the face of it, it's being necessary that water is H_2O is devastating for the idea that conceivability is a guide to possibility. Conceivability is a decidedly a priori affair, and you can only learn that water is H_2O via empirical investigation. So if you don't already know that water is H_2O, then you're going to find it conceivable that water isn't H_2O. And, of course, you're going to be wrong. But in Chalmers's framework, you can avoid this problem by being more careful. If you're sure to focus on a statement's primary intension, *and* positively conceive of a scenario of which that statement is true, *and* you conceive in an ideal way, then you can know that that statement describes a genuine possibility. Conceivability, in a specific sense, entails possibility, in another specific sense.

This form of modal rationalism may work for various ordinary modal claims. However, it plays directly into van Inwagen's hands when it comes to the interesting ones. The issue stems from the fact that we aren't ideal conceivers. Given this, we need to consider when our conceivings are good enough to justify our modal judgments. When it comes to ordinary modal claims—where matters are relatively simple, or we have background knowledge that helps us assess the particular scenario in question—perhaps we can often approximate ideal conceiving.[12] But it's implausible that we approximate the ideal when matters are unfamiliar and/or hugely complex, where our limited understanding and processing power make us vulnerable to significant error.

Chalmers responds to this concern by saying that there should be a presumption in favor of *p*'s ideal conceivability when *p* is prima facie conceivable (1996, 2010). But given our many cognitive limitations—we certainly aren't ideal conceivers generally—I see no reason to grant such a presumption. For starters, there are the many sources of cognitive error: e.g., ignoring relevant details because of ignorance or bias, substituting an easier conceiving task for the one in question, myside bias, and our general inclination to believe rather than suspend judgment, whatever our evidence. (For others, see Kahneman (2011).) Moreover, we just can't hold much information in our heads at one time. As a result, when matters get complex, it's hard to rule out the possibility that a scenario seems conceivable only because we can't entertain all its parts simultaneously. The appearance of possibility is generated by each component appearing possible, which we mistakenly attribute to the whole. Finally, it's often the case that something is prima facie conceivable due to ignorance. Where we don't know information that's relevant to whether the scenario is genuinely possible, conceiving more carefully won't help: it will still seem to us that *p* is possible, but for the wrong reason.

I should expand on this last point. It seems to me that I can positively conceive of an invisible mug. This mug is a microphysical duplicate of the mug I'm actually using, but it's entirely transparent. Such an object is positively conceivable: it isn't that the scenario I'm imagining *fails to rule out* the existence of such a mug; it's a scenario that *includes* such a mug. However, I don't take this to be any guide to what's ideally conceivable. The reason for this is simple: if I could actually entertain every relevant detail about the microstructural and nomic facts, I doubt that I'd find it conceivable. In particular, I doubt that I'd be able to see how light was failing to be reflected by the surface of this mug, even though I have no trouble conceiving of it not being so reflected. I suspect that my ignorance of these details is what makes positive conceiving possible, and also no guide to possibility.

The temptation is to reply that the laws could be different. If they were, then light wouldn't necessarily be reflected by an object with the same microphysical properties as the one I possess. But *if* we already know that reflectance properties are distinct from other properties, *then* we can vary the laws to get different relations between them. However, the question here is whether reflectance properties *are*

[12] *Can*—not *do*. The point is just that we have the capacity, not that we often realize it.

distinct from other properties, or whether they're somehow reducible to them. So, even if we know that the laws could be different—about which I'm skeptical, for reasons already given—it would beg the question to invoke different laws at this juncture.

Given all the above, we shouldn't posit a presumption in favor of *p*'s ideal conceivability when *p* is prima facie conceivable. The alternative is akin to positing a presumption in favor of how small objects appear visually at great distances. We know that we make mistakes when we endorse such seemings, so we generally avoid endorsing them. Rather, we suspend judgment until we can take a closer look. Likewise, we know that prima facie conceivability doesn't entail possibility, which means that we make mistakes there too. What's more, our epistemic situation is *worse* when it comes to conceiving. Unlike visual perception, where we can use other sensory modalities to confirm or disconfirm our judgments, we have few tools with which to assess modal matters. This should make us more cautious, not less, as we're less likely to be corrected should we fall into error.

Instead, then, it seems more reasonable to say that there are circumstances in which we can make the case that prima facie conceiving approximates ideal conceiving, and there it should be trusted. Usually, these will be cases that are relatively simple, or where we have plenty of background knowledge that lowers the odds of conceiving out of ignorance. In other words, these will be cases where we're considering ordinary modal claims. But when considering extraordinary modal claims—such as a microphysical duplicate of me that lacks phenomenal consciousness—we won't be able to argue that prima facie conceiving approximates ideal conceiving. Hence, we shouldn't make the inference from prima facie conceivability—or secunda facie conceivability—to genuine possibility.

The upshot? Whatever the virtues of Chalmers's modal rationalism, it's hardly clear that it supports our justifiably believing interesting modal claims *sans* knowing the relevant theories.

6.3 Williamson

Timothy Williamson offers a different strategy for salvaging conceivability. Williamson's story is straightforward. First, he notes that counterfactuals play an important role in our cognitive economy, and that we're fairly good at assessing their truth values. That this is so should be expected:

> Our overall capacity for somewhat reliable thought about counterfactual possibilities is hardly surprising, for we cannot know in advance exactly which possibilities are or will be actual. We need to make contingency plans. In practice, the only way for us to be cognitively equipped to deal with the actual is by being cognitively equipped to deal with a wide variety of contingencies, most of them counterfactual (2007, 137).

Williamson then points out that you can define necessity and possibility in terms of the counterfactual conditional: a proposition is necessary if and only if, were its negation to be the case, then a contradiction would obtain; a proposition is possible if and only if it *isn't* the case that, if it were so, then a contradiction would be true. In short:

$$\Box A \equiv (\sim A \Box\!\!\rightarrow \bot)$$

$$\Diamond A \equiv \sim (A \Box\!\!\rightarrow \bot)$$

Williamson is then free to treat the epistemology of metaphysical modality as a special case of the epistemology of counterfactuals. Typically, we evaluate ordinary counterfactuals in the imagination. Suppose that I left a book in my son's room, where he is now sleeping. I try to determine whether, if I were to sneak in, I would be able to retrieve the book without waking him. I develop the antecedent of this counterfactual in my imagination: my mind turns to the squeaky floorboards, the locations of which I have never been good at remembering, and I recall past failures. The prospects of success look dim, and I judge that I would not retrieve the book without drama. Williamson's claim is that roughly the same process can provide insights about metaphysical modality. The aim there, however, is to develop the antecedent in the interest of ferreting out contradictions, if any there be. As he puts it,

> ...we assert □A when our counterfactual development of the supposition ~A robustly yields a contradiction; we deny □A when our counterfactual development of ~A does not robustly yield a contradiction (and we do not attribute the failure to a defect in our search). Similarly, [...] we assert ◊A when our counterfactual development of the supposition A does not robustly yield a contradiction (and we do not attribute the failure to a defect in our search); we deny ◊A when our counterfactual development of A robustly yields a contradiction. Thus our fallible imaginative evaluation of counterfactuals has a conceivability test for possibility and an inconceivability test for impossibility built in as fallible special cases (2007, 163).

Like Yablo, Williamson uses the imagination to underwrite conceivability as a test for possibility. Williamson, however, makes no appeal to the principle of credulity, and he doesn't seem to think that there is anything special about the imagination other than its being the means by which we assess counterfactuals. On his view, the heavy lifting seems to be done by our reliability as counterfactual reasoners. (It's unclear to me whether he takes belief in that reliability to be justified by evolutionary considerations, though he obviously hints in that direction.) The rest of the work is done by the logical equivalences—meant, of course, to show that the capacities that make us reliable about ordinary matters will also allow us to be reliable about matters further removed from the business of everyday life.

C. S. Jenkins (2008) complains that this is all too quick: perhaps we *could* come by our modal knowledge this way, but that doesn't show that we do; Williamson needs more than the logical equivalences (plus a general appeal to theoretical economy, heretofore unmentioned) to show that he's identified *our* source, and not just a potential one. But Jenkins is confused: it may well be that we don't often perform the imaginative exercises to which Williamson alludes, but no matter, since his theory doesn't predict otherwise. Roughly, what Williamson's theory predicts is this: if we justifiably believe that *p* is necessary, then were we to consider how things would be if *p* were false, we would detect a contradiction; analogously, if we justifiably believe that *p* is possible, then were we to consider how things would be if *p* were true, we would detect no contradiction, nor would we see anything wrong with our search for one.[13] Williamson's theory is plausible insofar as (a) these predictions are accurate and (b) their accuracy explains the justification that attaches to the relevant beliefs. He need not offer a theory of the genesis of our modal beliefs in order to offer a theory of their justification.

I think that these predictions probably *are* accurate, but their accuracy doesn't explain the justification that attaches to the relevant beliefs. Rather, they're accurate because Williamson helps himself to some assumptions that do the heavy lifting in his epistemology—not the counterfactual framework that he develops.

We can see this by revisiting some familiar terrain. Kripke convinced most of us that there are necessary truths that can only be known a posteriori: necessarily, water is H_2O, heat is mean molecular kinetic energy, lightning is electrical discharge, and so on.[14] Williamson grants that there are such truths, but he contends that they pose no problem for his epistemology; they simply require that we be appropriately well-informed:

> If we know enough chemistry, our counterfactual development of the supposition that gold is [not] the element with atomic number 79 will generate a contradiction. The reason is not simply that we know that gold is the element with atomic number 79, for we can and must vary some items of our knowledge under counterfactual suppositions. Rather, part of the general way we develop counterfactual suppositions is to hold such constitutive facts fixed (2007, 164).

Essentially, Williamson is contending that *noseeum* inferences underwrite possibility claims: we don't see any contradictions, so they ain't there. As he fully appreciates, the quality of a noseeum inference is only as good as the degree to which our abilities fit the subject matter at hand. Hence, he stresses the importance of having the relevant chemical knowledge. But ours is a world with a serious division of cognitive labor; unless you're a specialist on *x*, then you probably have little more than rudimentary knowledge of the 'constitutive facts' about *x*. One

---

[13] I'm not completely confident whether the language of 'justified belief' is most appropriate here; Williamson would probably prefer 'knowledge'. The important point for my purposes, however, is the structure of the prediction, not the precise epistemic vocabulary.

[14] You can resist Kripke's conclusion by denying that there are genuine proper names on both sides of the identity sign, maintaining instead that at least one of them is a definite description in disguise; then, you can make a scope distinction to avoid having the necessity attach to the properties ascribed by the definite description. This shows that you can resist the conclusion; it does not show that you should. Whether you should turns on how plausible it is that we use certain terms—e.g., 'H_2O'—as proper names. If we do, then we're committed to the necessity; if we don't, then we aren't. Most of us think that we do. You might ask whether we ought to use those terms as proper names. I don't know how to answer this question.

unfortunate thing about having rudimentary knowledge is that you often don't know what inferences to make based on the knowledge you've got: I know, for example, that water is H_2O, but I know very little about the implications of this fact for water's behavior in various circumstances. So, if Williamson's account is right, then I'm probably not in a position to make many interesting modal claims about water, or about anything else beyond my expertise; I ought to limit myself to inferences from actuality to possibility, plus claims licensed by testimony from experts. And this is precisely what modal skepticism says. So, even if there aren't any problems with Williamson's account, it's still a cautious modal epistemology. For example, since I'm not up on my metallurgy, I couldn't use Williamson's account to say whether there could be transparent iron.

To be fair, Williamson acknowledges this:

> A nuanced account of our handling of counterfactuals is likely to predict that we are more reliable in evaluating some kinds than others. For example, we may well be more reliable in evaluating counterfactuals whose antecedents involve small departures from the actual world than in evaluating those whose antecedents involve much larger departures. We may be correspondingly more reliable in evaluating the possibility of everyday scenarios than of "far-out" ones, and extra caution may be called for in the latter case. At the limit, actuality is often the best argument for possibility. But current philosophical practice already shows some sensitivity to such considerations. Many philosophers are more confident in their judgments about more or less realistic thought experiments in epistemology and moral philosophy than about more radically strange ones in metaphysics (2007, 164).

However, it isn't clear that this concession goes far enough.

Williamson's account is either (a) incomplete or (b) too permissive. Consider this: why can't we imagine how things would be if gold were the element with atomic number 78? It certainly *seems* as though we can hold fixed the object of our imagining—in this case, gold—and this whether or not we also hold fixed its being the element with atomic number 79. Williamson blocks this by saying that we're supposed to hold *constitutive* facts fixed. But why, exactly? If we know that we should, then *of course* we'll be able to derive contradictions, since the antecedent will always (implicitly or explicitly) contradict the consequent: e.g., "If gold—i.e., the element with atomic number 79—were the element with atomic number 78, then a contradiction would be true." But then it's our knowledge of this obligation that's of interest to the modal epistemologist, and not the machinations in the counterfactual imagination. How do we come by this knowledge? Is it a priori? Does it fall out of our concept of a thing that has a constitution? Is it gleaned from some abductive argument? In the absence of answers to these questions, Williamson's doesn't actually have a modal epistemology for *any* modal claims, not just interesting ones. Instead, he has a schematic that could be fleshed out in any number of ways. Alternately, suppose that we *aren't* required to hold such constitutive facts fixed. Then whence the contradiction? Now it looks like his account is far too permissive: what *won't* we judge to be possible under this interpretation? (Note that it won't do to appeal here to the necessity of identity, at least if that principle and it's application here, via various theses in the philosophy of language, are motivated by the same intuitions that lead Williamson to say that we should hold constitutive facts fixed. Since those intuitions are in question, an argument for their veracity is wanted before we take them seriously.)

Williamson's view, I think, is that his counterfactual account is incomplete, and in more recent work—e.g., Williamson (forthcoming)—he makes some suggestions about how we can assess which properties are essential (and so should be held fixed) and which can be varied. His story there is abductive: we justify judgments about the essential properties of entities by appealing to their roles in our best explanatory frameworks, which is supposed to be a process that's continuous with what occurs in the natural sciences. But that view is obviously akin to TEM, and unlikely to be much more permissive. In any case, absent such an account, what Williamson provides is a schema that could be filled out in a number of ways. (Indeed, you might even regard TEM as one such way. According to TEM, the essential properties are the ones that are invariant in the theory of the target system; the contingent ones are the rest, at least by the lights of that theory.) And with such an account, Williamson's view remains unlikely to underwrite knowledge of interesting modal claims.

6.4 Conclusion

We've now considered a variety of ways to underwrite justification concerning interesting modal matters. None succeeds. So, although TEM is a cautious modal epistemology, and in this sense won't be judged to be conservative by the lights of optimists, that conservatism shouldn't count against it. The inclination to say that we have interesting modal justification is too difficult to support.

Chapter 7
TEM and the Theoretical Virtues

Abstract This concluding chapter summarizes the case for TEM and revisits a point with which I began: namely, that TEM needn't compete with every other modal epistemology; it can be a supplement, not a rival. To make this point clearer, I briefly discuss one essence-based modal epistemology, showing how TEM can deliver a key principle for that epistemology that might otherwise be difficult to defend.

TEM isn't a complete modal epistemology. TEM is a non-uniform account, allowing that some modal claims can be justified without explicit recourse to theory—namely, the ordinary ones, such as those concerning the possible locations of furniture in a room. However, it's no objection to scientific theorizing that we can answer some questions without doing science, and the same point applies here. Moreover, others have already proposed accounts of ordinary modal justification that could supplement TEM—see, e.g., Leon (2016) and Roca-Royes (2016)—and which fit neatly with the game analogy. (Leon (2016) defends the view that we can use induction and arguments from analogy to justify ordinary modal claims, and Roca-Royes (2016) defends arguments based on similarity for the same purpose.) I offer TEM as the standard to which to hold interesting modal claims—nothing more, nothing less.

I don't claim that we should believe TEM. To take that stance would be to ignore the arguments I offered concerning the criteria of theory selection. If TEM fares well on the virtues, then we should accept it in Dawes's sense: we should be willing to employ it as a premise in our reasoning, whether theoretical or practical, in any domain to which it might apply, with the goal of attaining knowledge. This is an appropriately cautious attitude to take toward an appropriately cautious modal epistemology.

Why accept TEM? The basic argument for it stems from the game analogy. We don't make up the modal facts and we don't have a faculty that acquaints us with them. So, just as we'd try to determine what's permissible and impermissible in a game with which we were unfamiliar, we should use perfectly ordinary tools to make modal judgments: we should make inferences from actuality to possibility, offer analogical and inductive arguments, and do our best to offer theories about the world's rules. Observation, analogy, and induction are often adequate for

B. Fischer, *Modal Justification via Theories*, Synthese Library 380,
DOI 10.1007/978-3-319-49127-1_7

ordinary modal judgments, but as we consider interesting modal claims—whether technological, scientific, or metaphysical—we should look to our best theories for guidance, just as TEM requires.

What's more, TEM fares well on plausible criteria for choosing among available modal epistemologies. First, it's useful for diagnosing modal disagreements, since it encourages us to use a familiar set of questions to locate, clarify, and ultimately resolve disagreements about modal claims by focusing on the theories behind them. We know how to disagree well about theories; TEM tells us to disagree the same way about interesting modal claims.

Second, TEM is a simple theory. It takes two unassuming ideas—that we justifiably believe our best theories, and that these theories have modal implications—and combines them into a modal epistemology. It doesn't require more from us than epistemic humility, which can be motivated on independent grounds (as I've argued).

Third, TEM is conservative. In part, this is because it fares no worse than other modal epistemologies in its account of the scope of our modal justification. But conservativism isn't just about what a theory says about particular modal claims: it's also about how it fits with our other convictions and practices. TEM's building blocks are familiar commitments for scientific realists. Likewise, its demand that we inspect our theoretical commitments fits with how we adjudicate modal disputes. It also fits with an intuitive hypothesis about the relationship between the scope of our knowledge of two domains—the more you know about actuality, the more you know modal reality—and it fits with a plausible idea about how we should discover facts that aren't of our making.

I want to conclude by returning to a point with which I began: namely, that TEM needn't compete with every other modal epistemology; it can be a supplement, not a rival.

To see what I mean, let's consider Hale's (2013) essence-based account. Hale thinks that knowledge of essence precedes modal knowledge, which commits him to offering an account of how we acquire knowledge of essence. Here is his summary:

> In a nutshell, [the strategy for acquiring knowledge of essence] consists in formulating and arguing for general principles of essence—principles asserting, schematically, that such-and-such a property is essential to its instances—from which we may infer specific Kripke conditionals [i.e., conditionals of the form, if p, then $\Box p$] which, in their turn, may serve as the major premises for Kripke-style inferences to specific essentialist conclusions [e.g., if Hesperus is Phosphorus, then necessarily, Hesperus is Phosphorus; Hesperus is Phosphorus; so, necessarily, Hesperus is Phosphorus] (269).

What matters, as Hale goes on to note, is whether we can defend specific general principles, such as the necessity of kind membership. If we can, then we've got our epistemology. If we can't, then we don't.

In a sense, Hale's view is already a version of TEM: he thinks that our modal knowledge depends on our best theories, and just happens to think that those theories concern the essences of objects. So, I have no objection to his method, though I'm skeptical of the specific arguments he offers for the general principles.[1] However, arguing as much would take us too far afield, so here I just want to point out that *if* those arguments fail, TEM can solve Hale's problem.

Let's focus on the example of kind membership, as Hale does. The kind membership principle says:

> Any object is essentially an object of a certain general kind (2013, 270).

Given TEM, this principle is easy to defend. Theories involve categorization schemes: they divvy up the world into objects, and group those objects in various ways. What's more, being justified in believing a theory requires integrating it with the other theories you justifiably believe. So, if you're justified in believing your total theory—i.e., the conjunction of all the theories that you justifiably believe—then that theory won't have inconsistent categorization schemes. Objects won't be material by the lights of one and immaterial by the lights of another. (Suppose otherwise. Then, you'd have a defeater for your total theory: no theory should ascribe inconsistent properties to objects.) But if that's right, then your total theory won't have models in which objects are of different general kinds. And *that* means that, according to TEM, you can be justified in believing that any object is essentially an object of a certain general kind. TEM can give Hale just the principle he wants.

I could make similar points about Peacocke's (1999) principles of possibility, Williamson's (2007) constitutive principles, Lowe's (2012) essence-based view, and many others. The challenge for them all is to explain how certain general theses are true, and necessarily so. TEM makes that project simple. Certain general theses are going to be true of all models specified by our total theory; hence, if we're justified in believing our total theory, we can be justified in believing that those general theses are necessary. Of course, there remain interesting questions about which principles those are, about the specific mental operations allow us to reason from and with those principles, and much else besides. So TEM too can be supplemented by theories that focus more squarely on those sorts of issues. The point is just that TEM needn't be a threat to several existing projects in modal epistemology: instead, it can be used as their foundation. Given the arguments of this book, that should be a welcome conclusion.

[1]For example, Hale's argument for the kind membership principle (given just below) is based on the idea that, to survive changes, a being has to remain the same F, where F is a "pure" sortal—e.g., "horse" instead of "brown horse." But since that sortal is one that applies whenever the object exists, and the object would cease to exist were that sortal *not* to apply, it seems plausible that sortal is essential to the object. And since this argument is perfectly general, it seems to support the kind membership principle. However, that argument only gets Hale *maximally general* sortal, such as "object"—it doesn't get him more specific sortals, like "horse." You can't defend the more *specific* kind membership principle, "A horse is essentially an object of the horse kind," without knowing that specific horses can't become other things. That's what Hale wants, but it isn't clear how he can get it.

References

Armstrong, D. M. (1989). *A combinatorial theory of possibility*. New York: Cambridge University Press.

Baker, L. R. (2007). *The metaphysics of everyday life: An essay in practical realism*. New York: Cambridge University Press.

Bealer, G. (2004). The origins of modal error. *Dialectica, 58*, 11–42.

Benacerraf, P. (1973). Mathematical truth. *The Journal of Philosophy, 70*, 661–679.

Benacerraf, P., & Putnam, H. (1983). Introduction. In P. Benacerraf & H. Putnam (Eds.), *Philosophy of mathematics: Selected readings* (pp. 1–40). New York: Cambridge University Press.

Biggs, S. (2010). Abduction and modality. *Philosophy and Phenomenological Research, 83*, 283–326.

Bueno, O., & Shalkowski, S. (2004). Modal realism and modal epistemology: A huge gap. In E. Weber & T. De Mey (Eds.), *Modal epistemology* (pp. 93–106). Brussels: Royal Flemish Academy of Belgium.

Chalmers, D. (1996). *The conscious mind: In search of a fundamental theory*. New York: Oxford University Press.

Chalmers, D. (2002). Does conceivability entail possibility? In T. Gendler & J. Hawthorne (Eds.), *Conceivability and possibility* (pp. 145–200). New York: Oxford University Press.

Chalmers, D. (2010). *The character of consciousness*. New York: Oxford University Press.

Churchland, P. S. (1986). *Neurophilosophy: Toward a unified science of the mind-brain*. Cambridge, MA: MIT Press.

Churchland, P. M. (1989). *A neurocomputational perspective: The nature of mind and the structure of science*. Cambridge, MA: MIT Press.

Cohnitz, D. (2002). Modal skepticism: Philosophical thought experiments and modal epistemology. *Vienna Circle Institute Yearbook, 10*, 281–296.

Dawes, G. (2013). Belief is not the issue: A defense of inference to the best explanation. *Ratio, 26*, 62–78.

Day, T., & Kincaid, H. (1994). Putting inference to the best explanation in its place. *Synthese, 98*, 271–295.

DePaul, M. R., & Ramsey, W. (1998). *Rethinking intuition: The psychology of intuition and its role in philosophical inquiry*. Lanham: Rowman & Littlefield.

Evnine, S. (2008). Modal epistemology: Our knowledge of necessity and possibility. *Philosophy Compass, 3*, 664–684.

Fischer, R. W. (2014). Why it doesn't matter whether the virtues are truth-conducive. *Synthese, 191*, 1059–1073.

B. Fischer, *Modal Justification via Theories*, Synthese Library 380,
DOI 10.1007/978-3-319-49127-1

Fischer, R. W. (2015). Theory selection in modal epistemology. *American Philosophical Quarterly, 52*, 289–304.
Fischer, B. (2016). Modal empiricism: Objection, reply, proposal. In B. Fischer & F. Leon (Eds.), *Modal epistemology after rationalism*. Dordrecht: Springer.
Fischer, B. (Forthcoming). *C. I. Lewis and the Benacerraf problem*. Episteme.
Fumerton, R. (1980). Induction and reasoning to the best explanation. *Philosophy of Science, 47*, 589–600.
Gale, R. (1996). Some difficulties in theistic treatments of evil. In D. Howard-Snyder (Ed.), *The evidential argument from evil* (pp. 206–218). Bloomington: Indiana University Press.
Geirsson, H. (2005). Conceivability and defeasible modal justification. *Philosophical Studies, 122*, 279–304.
Gentner, D., & Stevens, A. L. (1983). *Mental models, cognitive science*. Hillsdale: Erlbaum.
Giaquinto, M. (2008). The linguistic view of a priori knowledge. *Philosophy, 83*(1), 89–111.
Giere, R. N. (1979). *Understanding scientific theories*. New York: Holt, Rinehart, and Winston.
Gopnik, A., & Meltzoff, A. N. (1997). *Words, thoughts, and theories: Learning, development, and conceptual change*. Cambridge: MIT Press.
Gregory, D. (2010). Conceivability and apparent possibility. In B. Hale & A. Hoffmann (Eds.), *Modality: Metaphysics, logic, and epistemology* (pp. 319–342). New York: Oxford University Press.
Gunner, D. L. (1967). Professor Smart's 'Sensations and brain processes'. In C. F. Presley (Ed.), *The identity theory of mind* (pp. 1–20). St. Lucia: University of Queensland Press.
Hájek, P., & Pudlák, P. (1998). *Metamathematics of first-order arithmetic* (2nd ed.). Berlin: Springer.
Hale, B. (2003). Knowledge of possibility and necessity. *Proceedings of the Aristotelian Society, 103*, 1–20.
Hale, B. (2013). *Necessary beings: An essay on ontology, modality, and the relations between them*. New York: Oxford University Press.
Hanrahan, R. (2007). Imagination and possibility. *Philosophical Forum, 38*, 125–146.
Hanrahan, R. (2009). Getting God out of our (modal) business. *Sophia, 48*, 379–391.
Harman, G. (1986). *Change in view: Principles of reasoning*. Cambridge, MA: MIT Press.
Hart, W. D. (1988). *The engines of the soul*. New York: Cambridge University Press.
Hart, W. D. (2003). The music of modality. *Topoi, 22*, 135–142.
Hawke, P. (2011). Van Inwagen's modal skepticism. *Philosophical Studies, 153*, 351–364.
Huemer, M. (2001). *Skepticism and the veil of perception*. Lanham: Rowman & Littlefield.
Huggett, N. (2010). *Everywhere and everywhen: Adventures in physics and philosophy*. New York: Oxford University Press.
Hume, D. (1739). In L. A. Selby-Bigge & P. H. Nidditch (Eds.), *A treatise of human nature*. New York: Oxford University Press.
Jackson, F., Pargetter, R., & Prior, E. (1982). Functionalism and type-type identity theories. *Philosophical Studies, 42*, 209–225.
Jenkins, C. S. (2008). Modal knowledge, counterfactual knowledge and the role of experience. *Philosophical Quarterly, 58*, 693–701.
Johnson-Laird, P. N. (2004). The history of mental models. In K. Manktelow & M. C. Chung (Eds.), *Psychology of reasoning: Theoretical and historical perspectives* (pp. 179–212). New York: Psychology Press.
Kahneman, D. (2011). *Thinking, fast and slow*. New York: Farrar, Straus and Giroux.
Kallestrup, J. (2006). Physicalism, conceivability and strong necessities. *Synthese, 151*, 273–295.
Kelly, K. (2007). A new solution to the puzzle of simplicity. *Philosophy of Science, 74*, 561–573.
Kim, J. (1994). Explanatory knowledge and metaphysical dependence. *Philosophical Issues, 5*, 51–69.
Kind, A. (2001). Putting the image back in imagination. *Philosophy and Phenomenological Research, 62*, 103–105.
Klein, C. (2008). An ideal solution to disputes about multiply realized kinds. *Philosophical Studies, 140*, 161–177.

Koons, R. C. (2000). *Realism regained: An exact theory of causation, teleology, and the mind*. New York: Oxford University Press.
Kripke, S. (1980). *Naming and necessity*. Cambridge, MA: Harvard University Press.
Kung, P. (2010). Imagining as a guide to possibility. *Philosophy and Phenomenological Research, 81*, 620–663.
Kvanvig, J. (1989). Conservatism and its virtues. *Synthese, 79*, 143–163.
LaPorte, J. (2004). *Natural kinds and conceptual change*. New York: Cambridge University Press.
Leon, F. (2009). *Armchair knowledge and modal skepticism: A rapprochement*. Doctoral Dissertation, University of California, Riverside.
Leon, F. (2016). From modal skepticism to modal empiricism. In B. Fischer & F. Leon (Eds.), *Modal epistemology after rationalism*. Dordrecht: Springer.
Lewis, C. I. (1970). The pragmatic element in knowledge. In J. D. Goheen & J. L. Mothershead (Eds.), *Collected papers of Clarence Irving Lewis* (pp. 240–257). Stanford: Stanford University Press.
Lewis, D. (1986). *On the plurality of worlds*. Malden: Blackwell Publishers.
Lipton, P. (2004). *Inference to the best explanation*. New York: Routledge.
Lloyd, E. A. (1994). *The structure and confirmation of evolutionary theory*. Princeton: Princeton University Press.
Lowe, E. J. (2008). Two notions of being: Entity and essence. *Royal Institute of Philosophy Supplements, 83*, 23–48.
Lowe, E. J. (2012). What is the source of our knowledge of modal truths. *Mind, 121*, 919–950.
Lycan, W. (1988). *Judgment and justification*. New York: Cambridge University Press.
McCain, K. (2008). The virtues of epistemic conservatism. *Synthese, 164*, 185–200.
Mcleod, S. (2005). Modal epistemology. *Philosophical Books, 46*, 235–245.
Molnar, G., & Mumford, S. (2003). *Powers: A study in metaphysics*. New York: Oxford University Press.
Nichols, S. (2006). Imaginative blocks and impossibility: An essay in modal psychology. In S. Nichols (Ed.), *The architecture of the imagination* (pp. 237–256). New York: Oxford University Press.
Nolan, D. (1997). Quantitative parsimony. *British Journal for the Philosophy of Science, 48*, 329–343.
Nozick, R. (2003). *Invariances*. Cambridge, MA: Harvard University Press.
O'Connor, T. (2008). *Theism and ultimate explanation: The necessary shape of contingency*. Malden: Blackwell.
Oreste Fiocco, M. (2007). Conceivability, imagination and modal knowledge. *Philosophy and Phenomenological Research, 74*, 364–380.
Peacocke, C. (1985). Imagination, experience, and possibility: A Berkeleian view defended. In J. Foster & H. Robinson (Eds.), *Essays on Berkeley* (pp. 19–36). Oxford: Clarendon Press.
Peacocke, C. (1999). *Being known*. New York: Oxford University Press.
Plantinga, A. (1993). *Warrant and proper function*. New York: Oxford University Press.
Plantinga, A., & Tooley, M. (2008). *Knowledge of God*. New York: Blackwell.
Pruss, A. (2015). Possibility is not consistency. *Philosophical Studies, 172*, 2341–2348.
Putnam, H. (1979). *Mathematics, matter, and method: Philosophical papers*. New York: Cambridge University Press.
Quine, W. V. (1963). On simple theories of a complex world. *Synthese, 15*, 103–106.
Quine, W. V., & Ullian, J. S. (1978). *The web of belief*. New York: Random House.
Rasmussen, J. (2013). Continuity as a guide to possibility. *Australasian Journal of Philosophy, 92*, 525–538.
Rea, M. C. (2002). *World without design: The ontological consequences of naturalism*. New York: Oxford University Press.
Riggs, W. (2008). Epistemic risk and relativism. *Acta Analytica, 23*, 1–8.
Roca-Royes, S. (2007). *Mind independence and modal empiricism*. In Latin meeting in analytic philosophy genova, 20–22 September 2007, CEUR-WS Proceedings (pp. 117–135).

Roca-Royes, S. (2011). Modal knowledge and counterfactual knowledge. *Logique et analyse, 54*(216), 537–552.

Roca-Royes, S. (2016). Similarity and possibility: An epistemology of *de re* possibility for concrete entities. In B. Fischer & F. Leon (Eds.), *Modal epistemology after rationalism*. Dordrecht: Springer.

Salmon, W. C. (1989). *Four decades of scientific explanation*. Pittsburgh: University of Pittsburgh Press.

Salmon, N. (2007). *Content, cognition, and communication: Philosophical papers*. New York: Oxford University Press.

Shalkowski, S. (2010). IBE, GMR, and metaphysical projects. In B. Hale & A. Hoffmann (Eds.), *Modality: Metaphysics, logic, and epistemology* (pp. 167–187). New York: Oxford University Press.

Shoemaker, S. (1993). The first-person perspective. *Proceedings of the American Philosophical Association, 68*, 7–22.

Sidelle, A. (1989). *Necessity, essence, and individuation: A defense of conventionalism*. Ithaca: Cornell University Press.

Sober, E. (1981). The principle of parsimony. *British Journal for the Philosophy of Science, 32*, 145–156.

Sosa, E. (2000). Modal and other a priori epistemology: How can we know what is possible and what impossible? *The Southern Journal of Philosophy, 38*, 1–16.

Stich, S. P. (1990). *The fragmentation of reason: Preface to a pragmatic theory of cognitive evaluation*. Cambridge, MA: MIT Press.

Strawson, P. F. (1959). *Individuals: An essay in descriptive metaphysics*. London: Methuen.

Sturgeon, S. (2010). A priorism and modality. In B. Hale & A. Hoffmann (Eds.), *Modality: Metaphysics, logic, and epistemology* (pp. 275–310). New York: Oxford University Press.

Suppe, F. (1977). *The structure of scientific theories*. Urbana: University of Illinois Press.

Suppe, F. (1989). *The semantic conception of theories and scientific realism*. Chicago: University of Illinois Press.

Suppes, P. (1993). *Models and methods in the philosophy of science: Selected essays*. Dordrecht: Kluwer Academic.

Swinburne, R. (2001). *Epistemic justification*. New York: Oxford University Press.

Swinburne, R. (2004). *The existence of God*. New York: Oxford University Press.

Tahko, T. E. (2009). On the modal content of a posteriori necessities. *Theoria, 75*, 344–357.

Taliaferro, C. (1986). A modal argument for dualism. *Southern Journal of Philosophy, 24*, 95–108.

Thomasson, A. (2007a). Modal normativism and the methods of metaphysics. *Philosophical Topics, 35*, 135–160.

Thomasson, A. L. (2007b). *Ordinary objects*. New York: Oxford University Press.

Thompson, P. (1989). *The structure of biological theories*. New York: State University of New York Press.

Tuomela, R. (2000). Belief versus acceptance. *Philosophical Explorations, 3*, 122–137.

Vaidya, A. J. (2008). Modal rationalism and modal monism. *Erkenntnis, 68*, 191–212.

Vaidya, A. J. (2010). Understanding and essence. *Philosophia, 38*, 811–833.

Vaidya, A. (2015). The epistemology of modality. In E. N. Zalta (Ed.), *The stanford encyclopedia of philosophy*. http://plato.stanford.edu/archives/sum2015/entries/modality-epistemology

Vaidya, A. (2016). Modal knowledge: Beyond rationalism and empiricism. In B. Fischer & F. Leon (Eds.), *Modal epistemology after rationalism*. Dordrecht: Springer.

Van Fraassen, B. (1980). *The scientific image*. New York: Oxford University Press.

Van Fraassen, B. (1989). *Laws and symmetry*. New York: Oxford University Press.

Van Inwagen, P. (1977). Ontological arguments. *Nous, 11*, 375–395.

Van Inwagen, P. (1979). Review: The coherence of theism, by Richard Swinburne. *The Philosophical Review, 88*, 668–672.

Van Inwagen, P. (1995). *God, knowledge & mystery: Essays in philosophical theology*. Ithaca: Cornell University Press.

Van Inwagen, P. (1997). Materialism and the psychological-continuity account of personal identity. *Philosophical Perspectives, 11*, 305–319.
Van Inwagen, P. (1998). Modal epistemology. *Philosophical Studies, 92*, 67–84.
Van Inwagen, P. (2006). *The problem of evil: The Gifford lectures delivered in the University of St. Andrews in 2003*. New York: Oxford University Press.
van Inwagen, P. (2009). *Metaphysics* (3rd ed.). Boulder: Westview.
Vogel, J. (1990). Cartesian skepticism and inference to the best explanation. *The Journal of Philosophy, 87*(11), 658–666.
Weisberg, M. (2006). Water is not H_2O. In D. Baird, E. Scerri, & M. I. Lee (Eds.), *Philosophy of chemistry* (pp. 337–345). Netherlands: Springer.
Williamson, T. (2007). *The philosophy of philosophy*. Malden: Blackwell.
Williamson, T. (Forthcoming). Modal science. In A. Vaidya & D. Prelević (Eds.), *The philosophy of modality: The standard model and beyond*. New York: Oxford University Press.
Wright, C. (1986). Inventing logical necessity. In J. Butterfield (Ed.), *Language, mind and logic* (pp. 187–210). Cambridge, MA: Cambridge University Press.
Yablo, S. (1990). The real distinction between mind and body. *Canadian Journal of Philosophy, 20*, 149–201.
Yablo, S. (1993). Is conceivability a guide to possibility? *Philosophy and Phenomenological Research, 53*, 1–42.
Yablo, S. (2006). No fool's cold: Notes on illusions of possibility. In M. Garcia-Carpintero & J. Macia (Eds.), *Two dimensional semantics* (pp. 327–345). New York: Oxford University Press.
Yandell, K. (1988). Divine necessity and divine goodness. In T. V. Morris (Ed.), *Divine and human action: Essays in the metaphysics of theism* (pp. 313–344). Ithaca: Cornell University Press.

FSC
www.fsc.org
MIX
Papier | Fördert gute Waldnutzung
FSC® C083411

Kolibri 360 GmbH
Ferdinand-Jühlke-Straße 7
99095 Erfurt, Deutschland
produktsicherheit@kolibri360.de